Gorillas in Our Midst

**A Zookeeper's Tale
of Hand-Rearing
Baby Gorillas**

Alan Toyne

summersdale

GORILLAS IN OUR MIDST

Copyright © Alan Toyne, 2025

All rights reserved.

Photo of Alan and Hasani © Imogen Calendar; all other photos © Miriam Haas; gorilla drawing on pp.14, 103 and 220 © Natata/Shutterstock.com

No part of this book may be reproduced by any means, nor transmitted, nor translated into a machine language, without the written permission of the publishers.

Alan Toyne has asserted their right to be identified as the author of this work in accordance with sections 77 and 78 of the Copyright, Designs and Patents Act 1988.

Condition of Sale
This book is sold subject to the condition that it shall not, by way of trade or otherwise, be lent, resold, hired out or otherwise circulated in any form of binding or cover other than that in which it is published and without a similar condition including this condition being imposed on the subsequent purchaser.

An Hachette UK Company
www.hachette.co.uk

Summersdale Publishers
Part of Octopus Publishing Group Limited
Carmelite House
50 Victoria Embankment
LONDON
EC4Y 0DZ
UK

www.summersdale.com

This FSC® label means that materials used for the product have been responsibly sourced

MIX
Paper | Supporting responsible forestry
FSC® C104740

The authorized representative in the EEA is Hachette Ireland, 8 Castlecourt Centre, Dublin 15, D15 XTP3, Ireland (email: info@hbgi.ie)

Printed and bound in the UK

ISBN: 978-1-83799-525-7

Substantial discounts on bulk quantities of Summersdale books are available to corporations, professional associations and other organizations. For details contact general enquiries: telephone: +44 (0) 1243 771107 or email: enquiries@summersdale.com.

Author's Note

This is an account of my time at Bristol Zoo and all events relating to the gorillas are true. Both Hannah and Bob, however, are composite characters as firstly I didn't feel comfortable putting words into the mouths of my actual colleagues, and secondly the cast of characters is already enormous with all the gorillas. I've supplied two family trees, on pages 54 and 221, to help you get a sense of who is who in the troop as we hand-reared the two baby gorillas.

Prologue

Can You Get a Gorilla into a Car Seat?

The gorilla (a baby about the size of a rugby ball at seven months old) lies on your colleague's knee asleep. She affects a smile, beneath rubbery flared nostrils, smooth ridges of dark skin across her muzzle, long eyelashes and crazy wispy eyebrows. Her eyes flit behind closed lids, hands semi-clenched and ready in loose fists, thumbs relaxed, a brush of black hair across her knuckles. Legs bent, glossy fur trousers that end at her smooth-soled feet, toes half-clenched too, just like her fingers. Opposable big toes curled and relaxed, like her thumbs, with tiny smooth nails.

The overnight kit is ready: sterilized teething rings, nappy bags, the lion-headed baby toy that once made a roaring noise, nappies, bottles and more nappy bags. Milk powder and a perfectly ordered clipboard with all the right data recorded: sleep; feeding; poo; wee. Two pens, working, all fitted neatly in the 'hand-rearing bag', ready and waiting by the door, a box of disposable gloves on top.

You slide off your string vest, freshly laundered but worn-in for a couple of hours so it smells reassuringly familiar;

string vests are an essential part of the hand-rearing outfit. It is worn to replicate fur, and the baby gorilla learns to hang off it, knowing that when you move, it has to cling on to come with you, so it never gets left behind.

Your colleague lowers the gorilla into the car seat and you bundle the string vest in with the sleeping infant. Her fingers and toes twitch and hold the vest gently, she stays asleep as you clip her in. Sneak out of the Monkey Jungle office, creep down the metal steps and by some miracle the rowdy gang of spider monkeys don't notice as you exit the zoo through a side door behind their enclosure. The gorilla, snug in the baby chair, is belted into the front passenger seat and the vibration of the engine keeps her asleep until you get home.

That's how it *should* go, but usually…

Hannah stumbles into the office. Her hair has strands of wood-wool in it, her face is glazed and pale from a full two-night stretch of hand-rearing. Her string vest is clumped with baby gorilla shit and she honks of piss. Afia, the gorilla, is wide-eyed and awake, wired, hanging from Hannah's string vest with a big toe hooked onto her belt.

'Thank god that day's over,' Hannah says.

We've been roping the gorilla house today, splicing miniature Afia-sized ropes and ladders.

'Hey you,' I say to Afia. 'Have you worn Hannah out?'

'Two nights in a row is…' Hannah yawns. 'Well, you know what it's like.'

Our office overlooks the ring-tailed lemur enclosure, with its palm trees and climbing-frame of long branches, a walkway below for the public and a waterfall that trickles over smooth pebbles. A lemur alarm-calls outside, a high-pitched shriek-bark, which sets all the rest off in an echoing cacophony. They run along the paths, leap about and crash in and out of the battered palm trees outside. Our window is smeared with their scent markings.

Afia, transfixed, pulls herself up to look over Hannah's shoulder through the window. She frowns at the greenery and lemur chaos outside, takes it all in, hangs still.

'We need more data sheets,' Hannah says and the chair creaks as she flops into it.

Afia clambers higher up Hannah's back, one wrinkled hand on top of her head, fingers clamped in her hair. She cranes her neck to keep watch on the lemurs outside.

Through the tail-end of another yawn Hannah says, 'I dropped the clipboard in the drain… got all water-logged.'

I sit at the second wonky desk, stacked with a tower of over-stuffed plastic in-trays, and move an empty mug of dark, gleaming coffee residue out of Afia's grabbing range. The bowed shelves above the ancient computer are weighed down with animal diaries and spare UV lamp bulbs.

Afia loves the office.

'Come here, you.' I swivel my chair round, holding out my arms, and Hannah hands her over. Afia forgets the lemurs and stands up on my thighs as I hold her hands. Her sturdy little legs wobble as her toes grip my trousers. She has her 'play face' on, a big, toothy, open-mouthed smile.

Hannah leans back in her chair. 'She hasn't slept this afternoon.'

'Too busy playing with Bob,' I say to Afia. I let go of her hand to tickle her neck, inside her collar bone, and she dissolves into dry chuckles and tries to play-bite my thumb.

'And she tore the corner off the data sheet and tried to eat it.' Hannah holds up the clipboard and mutters, 'Two-hundred and eighty mils at the last feed,' to herself.

I know the look Hannah has on her face right now. I'm looking at the version of myself tomorrow, functioning on a level of immediate need and nothing more. She's on exhausted automatic, each task done methodically so as not to mess it up. The pain and misery of rectifying a mistake, losing those precious minutes, has ingrained a mode of functionality in us all, so once you sign off, nothing will disrupt that glorious uninterrupted sleep. Proper sleep in an actual bed. Without a gorilla.

During the day we sleep when Afia does. I stepped over them both at midday, Hannah out cold on the concrete floor of the gorilla introduction den, Afia slumped on top of her, little hairy arms around her neck, soft head wedged under her chin, legs wrapped around her ribs. The warmth of Afia is incredibly soporific, but you know any sleep you get will eventually be interrupted by a steady stream of hot gorilla piss that quickly chills as it soaks through your polo shirt.

'She's got a new tooth coming through too.' Hannah brings her computer to life. 'Upper number seven I think.'

I swivel back to the desk, letting Afia slap enthusiastically at my keyboard rather than have her bother Hannah as she tries to add the new tooth onto the daily report.

Afia grabs the phone with her little fingers and drops the receiver on the floor. Tugs the coiled spring-cord tight with both hands and a foot and jams it in her mouth to chew on.

'Go home,' I say to Hannah as I disentangle the cord from Afia's teeth. 'No way she's getting into the car seat for a while. I'll get someone else to give me a hand.'

'Thank you.' Hannah creaks round in her chair and almost bursts into tears. 'Thanks so much.' She tickles the back of Afia's head, her arm coated, like mine, in yellow and purple bite-sized circular bruises. 'Big day tomorrow.'

I see the realization fall across Hannah's face that if all goes well with Romina tomorrow, she may have spent her last night with Afia. She rubs her eye, pulls a strand of wood-wool out of her eyebrow and turns back to the keyboard. 'I'll print you some more data sheets,' she says, 'then I'll go home.'

Romina is the gorilla we've trained to become Afia's surrogate mum and we've agreed to try the first mix tomorrow.

The printer goes into overdrive and Afia clings to me in sudden terror. Phone cord forgotten, she grips me tight with hands and feet, pulls my string vest as she scurries around me away from the printer, eyes wide. She sucks her finger and ducks as the data sheets growl and jerk their way into a neat pile.

A ring-tailed lemur thumps onto the windowsill outside the office.

Afia jumps again, clambering right around me on the chair to cling onto my back, ready for us to flee. The exact response we've been aiming to instil in her these long months.

The ring-tail, grey fur down its back ruffled by the wind outside, peers in at us with beady orange eyes. The white pointed ears flick and its long, wet nose twitches. The eyes, ringed by black fur, are unable to comprehend the view of us through the window: two zookeepers in string vests, one baby gorilla. It turns its back on us, distracted as others alarm-call in the bushes.

Afia stretches her neck again to watch over my shoulder and I take advantage of her distraction, getting up to check the hand-rearing bag. She has her feet in the armpits of my vest, hands on my shoulders. I know the lemur is going to scent mark and I'll have time to go through the whole bag before Afia loses interest. I check all the bottle lids are screwed on tight.

The ring-tail leans forward on the windowsill, braces its spindly back feet against the glass, the miniature padded back toes splayed, flicks its striped tail aside and rubs its anal scent gland up and down the window.

Afia squirms higher still, arms around my neck, toes clenched. Our ears brush as she keeps the lemur in sight, moving her head as I stand at the window to re-pack the bag. Afia watches and absorbs what's going on, associates the noises with the movement and animals outside. I feel her limbs relax and she clambers back around me again. The chair creaks as we sit back down and she eyes the stapler while opening a drawer with her toes. A few more slaps at the keyboard and a frown as she sees the car seat waiting by the door, ready to take her out of the zoo and back to a house overnight.

It takes time to get her in, four attempts, interspersed with walks around the office, a trip around the empty zoo, nowhere near the gorilla house, not now that Afia has left for the day and is wearing a nappy. We can't have the rest of the troop seeing her wearing what they would deem to be clothes. They have to think of her as one of them, not a miniature hairy version of one of us. We pass the mara, meerkats and the squirrel monkeys, and she gazes at it all, nostrils twitching, with occasional halts to grab at the shrubs and bamboo.

Eventually she calms down enough to let me buckle her into the dreaded car seat back at the office and I give her my string vest to hold. The drive home is busy with traffic and Afia is beginning to look about the car and peer out of the window. I queue behind a four-by-four with two dogs in the back and stroke Afia's arm as we edge slowly down the hill. She pulls at the shoulder straps, hates being constrained and makes her little grunt-cough noise. A sound of irritation. If you were to snarl '*It*' but dropped the *t*. That's exactly the sound an annoyed baby gorilla makes.

Afia tugs at the straps some more, grunt-coughs and arches her back. Her lips push out, purse together and wobble – she's about to hoot, the *who, who, who* call that announces to the group that she's in trouble and needs rescuing.

I gorilla grumble in return – a deep rumble, like I'm expressing my delight at the most amazing roast dinner at Christmas – and nudge the car down the hill.

She tries a low *'who, who'* and really arches her back. Goes for it, kicks her legs, toes reaching for something to hold. Discards my string vest into the footwell.

I flick on the radio, and she instantly reaches for the lights on the display, remembers she's got her teething chew clamped in one fist and jams it into the side of her mouth, looking up at me.

'Won't be long,' I say and do another gorilla grumble. 'We'll be back at home, have a play on the sofa, eat some food, nice long sleep, long all-night kind of sleep.'

She chews the teething ring and kicks her legs, eyes drifting around the inside of the car again.

I make it through the traffic lights and onto the flyover, a twin line of brake lights up ahead. I see the dogs in the back of the four-by-four ahead pant and shuffle about.

Afia throws her teething chew into the footwell. Squirms, strains against the car seat straps. Her legs heave and she screams. No warning, no slow build-up of hoots, just high-pitched fear-screams. Repeatedly, she yanks and thrashes against the shoulder straps.

I'm on the inside lane next to a van, with no hard shoulder to stop on.

Afia shrieks.

There is a guy in the passenger seat of the van next to me, absorbed in his phone, with his headphones in and so he misses Afia force her way out of the car seat, slide out of the straps and use the steering wheel as a handhold to wedge herself onto my lap. She turns on the indicator as she clambers across.

'Pack it in.' I unzip my hoody, try and wrap it around her.

She clutches the gear stick with her foot and squirms around to face the steering wheel, tries to reach through it to turn the indicator back on.

The two dogs in front catch sight of her and go berserk.

Afia peers over the top of the steering wheel to stare at them.

The four-by-four rocks and we hear the bay of their muffled barks.

She yanks the screen-wash stalk, then jumps as blue liquid whines and sprays up the windscreen; spins to clutch me in terror as the wipers slash into life. I feel shit reverberate in her nappy and she wedges her head under my arm, quiet and momentarily wary.

She can't come home again.

That's the reality. Safety issues aside, the distress caused by the car seat is a result of being separated from me, exactly the kind of behaviour we have been trying to instil in her: that she must always cling on to her surrogate. Behaviour that is essential for her safety and development which will allow us to reintroduce her back with the rest of the gorillas. We are now going to have to live on site with Afia at the zoo, for as long as it takes to embed her back into the troop. This is the final time she'll come home, to spend the night with me and my family troop. Afia, born by emergency caesarean section, is now seven months old and ready to become a western lowland gorilla.

The path we're on has been trialled with mixed success a couple of times in Europe but never before in a UK zoo.

The realization throws the reality of what we're about to attempt into clarity. If all goes well tomorrow, Romina will pick Afia up and treat her as her own. I don't want to think about what could happen if it goes badly.

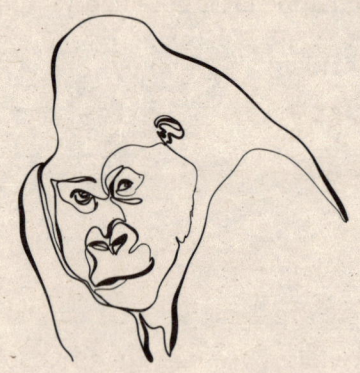

Part One
Life in the Troop

Chapter 1

First Day of Volunteering

My partner Sparky drops me off part way along the school run. She pulls into a layby with a slightly harried look and I climb out of the car, with my bag and packed lunch. The kids, Sam and Fizz, both wave out of the back window as she pulls away, gutted they must go to school and I'm off to the zoo for my first day as a volunteer keeper.

It's a 20-minute walk to the zoo across Clifton suspension bridge, which spans the Avon Gorge. The bridge is strung between two mammoth towers of precise curved bricks and taut cables. I don't get to walk this way often and stop halfway to gaze out across the city. I'm nervous. Excited.

Grand Edwardian terraced houses curve up the side of the gorge, with big windows and wrought-iron balconies. Where they can, trees hug the cliffs between shoulders of rock. It's low tide, the river a distant stream below, between banks of glistening mud.

I continue across the bridge, following little brown signs for the zoo and walk the broad, raised pavements through the streets of Clifton. This is the posh bit of town, where you take your partner's parents for coffee when they visit. Towering houses, most split into flats for rich students. Multitudes of black wheelie bins and BMWs parked all over the place.

It makes me happy that the brown 'Zoo this way' signs display a silhouette of a gorilla. For me it was – always – all about working with primates. I'd travelled when I left school, backpacked in the nineties, crashing through the jungles of Southeast Asia in search of orangutans, exploring the flooded forests of the Amazon trying to spot the bald uakari monkey, with their shaggy brown coats and bald crimson heads. Later I did an anthropology degree and became enthralled with primate social politics.

That's why Bristol Zoo is so perfect – it has specialized in the western lowland gorilla for 60 years and is home to a whole range of other primate species, including the aye-aye, those hairy, bat-eared nocturnal lemurs with the long, extendable fingers. The nocturnal house is also home to grey mouse lemurs, which weigh in at 200 grams. Jock, the zoo's silverback male gorilla, weighs 200 kilos and yet they, and we, as a species of primate, all have plenty in common. I'm fascinated by the similarities between other primate species and us humans, both physically and socially. How much of our behaviour is driven by inherited instinct from our shared, distant common ancestor?

The perimeter wall of the zoo is topped with a fence which overhangs the pavement and holds back spiky bushes. The zoo nestles behind Clifton Downs, between more fancy terraces and a big private school that the keepers would all refer to as Hogwarts.

A low repeated grunt echoes over the wall as I head through the tree-lined car park, from what sort of beast I can't tell, but then I do recognize the sudden whoop of gibbons.

I wait at the pair of tall double gates. I know I'm in the right spot, as I've checked the email obsessively. As promised in the directions, there is a small hobbit-sized door built into one of the giant gates.

It's closed.

I'm early, so I try to relax. I'm walking about with an internal glow. Getting a place as a volunteer has taken months and is a major step along the path to becoming a zookeeper.

The hobbit-door opens and a young guy pokes his head out, black spiky hair and stubble.

'Alright mate?' he says. 'You our new volley?'

'For mammals?'

'Come on in.' He holds the door open and I half crouch to step through it and into the zoo.

'I'm Al.' I hold out my hand.

He crunches it in a hefty handshake. Not an intentional macho bone-crusher, but a handshake of someone who works outside, does physical labour; calloused, firm and strong.

'Bob,' he says and grins. 'Welcome to the mad house.'

The little door clunks shut behind me.

'This here is stores.' Bob waves an arm at a long two-storey brick building. Functional, with big windows in faded aluminium frames. 'Maintenance workshop and then all the vets are upstairs.'

We pass battered corrugated doors that shield a forklift and a weird little green milk float-type vehicle, plugged into the mains with a twisted orange cable. Two long rusty chest freezers sit along the wall with laminated signs that shout '*20KG BAGS MAX!!!*'

Yellow wheelie bins, the big ones on four wheels with the kick breaks, are lined up outside a kitchen. Zookeepers stand at metal tables inside with rows of cardboard kidney dishes laid out in front of them.

'Diet prep in there later,' Bob says over his shoulder and leads me to a second set of tall mesh gates.

A sign reads: '*This is a lion-proof gate. Keep closed when not in immediate use.*'

'Got some time this morning,' he says, 'so I'll give you the tour round. Get changed first, then I'll show you where everything is.'

'Great,' I say.

He wears tattered faded shorts, the side pockets bleach-stained. His boots are chewed up across the toe caps and he has a radio on his belt, black with an aerial and a button you hold down to talk into. His green polo shirt has a tear across the front; tufts of dark wiry hair poke out and he wears a fleece over the top, at least two sizes too small for him.

We step through a mesh door set into the lion-proof gates. 'Got to grab some screws later,' he says with a nod at the maintenance workshop. 'Remind me.'

I peer through the broad windows and see rows of work benches and machine tools, racks of MDF and ladders.

'Back of the aquarium,' Bob continues with a wave at a row of giant water tanks.

They tower above us, thick industrial plastic held in frames of riveted steel. Pipes and heavy-duty valves that drip. 'Aquarium used to be a bear-pit back in the day,' he adds.

'Have you worked here long?'

'Straight from school mate,' he says. 'Get me ten-year anniversary badge next week.'

I follow him under a bridge, through a curtain of heavy plastic flaps, and we enter the public part of the zoo, pass the Art Deco restaurant, the front webbed with twisted wisteria branches and cascades of blue flowers. Rows of wooden picnic benches squat beneath a taut canvas awning, surrounded by dripping hanging baskets. A seagull lumbers out onto the lawn as I follow Bob.

'Why you with us then, Al?' he says.

'I want to be a zookeeper,' I reply without hesitation.

He smiles. 'You're a bit older than the normal volleys we get, no offence.'

He's right. My decision to become a zookeeper was late in life, by which I mean I'm 36. 'Better late than never,' I say.

He leads me across a wide lawn, the grass damp and springy. 'You'll get lost for the first week or two,' he says, 'but you're never really lost if you can find your way here. The main lawn.'

The gibbons whoop and hoot, letting out their morning territorial duet calls.

'I did some stuff with gibbons when I was at uni,' I venture. 'Their pair-bonding social system.'

'You went to university?' Bob looks at me, baffled.

The gibbons let off another round of whoops, long notes that build into a frenzied crescendo, before tailing off again.

'What are you ASBO lunatics shouting about now?' Bob says.

'ASBO?'

He nods. 'Some Clifton posho complained about the gib-gibs waking them up. We have to shut 'em in three nights a week.'

'Really?'

'Yeah, thing is, whoever it was that complained can't be older than the zoo, can they? Hundred and seventy-two years old. What you bloody expect living near animals?'

He leads me up a grass slope and through a wooden garden gate, between Portakabins and an enclosure full of tumbling ferrets, with ramps and shelves, old boots and cardboard boxes.

Bob ignores them and stops outside a door with a security code pad. 'Changing rooms upstairs.'

'Great.'

'Your uniform ain't here yet, so we've got you some dodgy overalls to wear and you've only got wellies until your boots turn up next week. All in locker number seven.'

'Okay, no worries.'

'You'll be useful today,' he says. 'In disguise.'

The changing room is lined with dented lockers and piled wellington boots, a plastic bin overflows in the corner and wet weather coats hang on pegs, torn with the stuffing poking out.

I pull on the overalls, dark blue, with poppers up the front, stamp my feet into a pair of green wellies and clump back downstairs to join Bob.

'The grand tour,' he says and rubs his big leathery hands together. 'We'll start up at langurs.'

I follow him to the main entrance, the glass doors still closed. The gift shop is crammed full of stuffed animals on shelves, all peering out, waiting for the day to begin.

Bob looks at his watch. 'Got twenty minutes yet, before we open,' he says and turns to stare into an enclosure, glass fronted with chicken wire-type mesh high above and a bleak-looking concrete pole in the middle. Two tiers of pallet shelving surround the pole and shelter a huddle of big, bright orange monkeys. They glare at us and hug each other's knees.

'Don't see many of these about.' Bob nods at the orange monkeys. 'They'll chuck shit at you if you give 'em half a chance. We had to put that cover up.' He points at strips of green cracked Perspex above the windows. 'They was pissing out through the mesh.' He laughs. 'Golden showers for the public.'

One of the langurs has black fur, huddled in the middle of everyone else. 'Is that the male?' I say. 'The black one.'

'Just a different colour morph.' Bob grins. 'Thought you studied at university?' He turns from the monkeys and points. 'This—' he pauses— 'this is the top terrace. There're pictures of this full of people, black and white photos back when zookeepers had to prance about in caps and polished boots, like a bus conductor.'

The terrace runs almost the entire length of the zoo, lined with trees and neat flowerbeds. Glossed green benches, with curved wrought-iron legs.

'We're the fifth eldest zoo in the whole world,' Bob says with pride. 'Opened in 1836, and this is the first zoo to not be in a capital city.'

'Really?' I'd read this, done my research, but I don't want to interrupt Bob's flow.

'Monkey puzzle tree down by bats,' he says, 'show you that in a bit. Trees here that are older than the zoo itself, two hundred years some of 'em, but never mind that. Lions first, just over here.' He leads me toward a line of windows, smeared with dinner-plate-sized paw prints.

'Every fucker moans about how they should be out on the plains in Africa and they ain't got enough room.'

'Oh right.' The enclosure feels small, but I've already put the captivity debate to one side in my mind. I'm here to learn and really get stuck in, understand how modern zookeeping works.

'They're Asiatic lions for a start,' Bob continues. 'So why they should be in Africa I don't know, and it might look small, but it's bigger in there than you think.'

The enclosure is roofed over with chain-link mesh and has bushes in front of a waterfall, some battered trees, a series of moss-covered rocks and tall wooden platforms at the back, with straw beds underneath. I see a lion face watching us.

'Plenty of cover for them, see.' Bob leans on the barrier. 'That's what they like, getting out of sight. Probably the best Asiatic lion enclosure in the country.' He turns to look at me. 'And Twilight World; that's got to be the best nocturnal house in the whole of Europe.' I hear the pride in his tone again.

'You've got aye-aye, haven't you?' I rein in babbling on about primate social systems and my anthropology degree. I need to fit in here and join the new troop.

'Those long-fingered idiots,' he says. 'Yeah, you don't see many of them about neither.'

The male lion's head appears next to the female's, mane tousled with straw.

'These dopey sods wouldn't know what to do in the wild anyway.' Bob frowns. 'You'll get a bit of that being a keeper. Bunny-huggers telling you the enclosures are too small.'

'Animal rights people?' A couple of my friends disagree with zoos, think they are cruel and that animals shouldn't be kept in captivity. They were bewildered by my decision to become a keeper.

Bob nods. 'There ain't no wild left for half the species we got here.'

His radio crackles. *'Fourteen to twenty-five Bob.'*

He unclips it from his belt. 'Receiving.'

'We need to get eyes on J.P.'

Bob looks at me. 'Tour's postponed.' He lifts his radio and says, 'Copy that, I've got our secret weapon.'

We cut back across the main lawn and the restaurant lights wink on as we pass, illuminating the tables and counters inside. The zoo shrugs awake, in preparation for the public coming in. I can smell coffee as I follow Bob into Monkey Jungle.

A woman waits for us, hair in a ponytail. She hugs herself, shivers and stamps her feet in her wellies. She has a green zoo polo shirt on and shorts like Bob's, but newer. 'You've got my fleece on,' she says.

'Just warming it up for you.' Bob looks at me. 'Ungrateful, ain't she?'

'Hi,' she says to me. 'Hannah.'

'Al,' I say, and we sort of almost shake hands, but she's too cold to stop hugging herself, so we don't. Both Bob and Hannah must be ten years younger than me at least. Mid-twenties.

'Has he been giving you the tour?' she asks.

Bob creeps down to the far enclosure window. 'Where's this clown now?'

Hannah frowns. 'He doesn't mean it.'

'Mean what?' Bob says over his shoulder.

'You're always calling your animals idiots and clowns,' Hannah says. 'Some people—' she gives me an apologetic smile '—they might find that offensive.'

'He wants to be a keeper,' Bob responds. 'You know I'm only joking, don't you Al?'

'Fleece.' She holds out her hand.

Bob steps back from the glass and peels it off. 'Chilly, ain't it?'

'I'm going to get quarantine done,' she says. 'Get back to Gorillas so we can take Kera out before break.'

'Righto.' Bob ushers me back toward the restaurant. 'We don't want that bearded idiot J.P. seeing us together.'

'That's exactly what I'm talking about,' Hannah calls over her shoulder as she heads off toward the maintenance block.

'Bearded idiot?' I ask.

'De Brazza monkeys.' Bob grins. 'We've got two of them in the last enclosure there.' He jabs his thumb over his shoulder.

'Breeding pair. J.P.'s the male, can't miss him, he's got bright blue bollocks.'

'Bright blue?'

'He's been getting out and over the moat onto Gorilla Island. Keeps doing it and we can't figure out how.'

'Okay.' I glance about. The thought of animal escapes hadn't occurred to me yet.

'He won't do it if one of us is watching,' Bob says. 'But we got you with us today. And you ain't in uniform. In disguise. He don't know you.'

'You want me to spy on him?'

'You'll have to keep a low profile, he's a crafty bastard. We think he's slipping through the electric fence or maybe jumping from a tree outside.'

'Okay, sure.'

'I'll be back to shut him in before break, so they can take Kera out.'

'Okay,' I say again, unclear of what he's saying or who Kera is. 'What if he does escape?'

'Just make sure you see how he's getting out. I'll be back in a bit with a radio, so you can contact us.'

Bob heads off and I sidle into Monkey Jungle. Four big enclosures, two on either side of a wide public walkway.

I lurk behind a handy screen of interpretation signs and peer into the De Brazza monkey enclosure. It has tall glass walls and a climbing frame of branches, screwed together and hung with ropes. There are piles of logs on the woodchip floor and a heavy concrete water bowl. A hatch leads outside to a row of trees and another wonky climbing frame, half screened from Gorilla Island by tall bamboo.

J.P. spots me hiding behind the sign straight away, my cover immediately blown. He stands on a sawn tree-ring platform inside, halfway up the climbing frame. He has a bright orange eyebrow ridge, above serious, dark eyes. He threatens me through the glass with a forward bob of his head. The white fur around his nose and chin forms a pointed beard and his shoulders are stocky, legs covered in black glossy hair. The grey speckled fur down his back looks almost crimped and his tail is long and dark. He sits down on the platform and shows me his blue baggy scrotum.

I can see straight through the glass back wall of J.P.'s enclosure, across the moat to Gorilla Island, a rolling hill with a long, sturdy climbing structure of massive beams and poles, leading to a brick building on the far side. The gorilla house.

One of the interpretation signs I'm hiding behind tells me all about the gorilla enclosure and provides a map of the island.

The hill, I read, allows the gorillas a space to be out of sight of the visitors if they want to. The island is covered with a selection of edible plants, which allow the animals to graze throughout the day, forage as they would naturally do in the wild. A laminated extra sign flaps as the breeze tugs through Monkey Jungle, stuck onto the interpretation board with drawing pins. An A4 sheet that says not to panic if you see gorillas up at the top of the large precarious tree toward the far end of the island.

I peer over the sign, through J.P.'s enclosure and scan the island for this precarious tree but can't see it, or any gorillas.

J.P. glares at me.

I read that the house has four entrances outside onto the island, but from here I can only pick out two gorilla-sized square metal hatches, closed and high up in the wall of the brick building on the far side of the island, one accessed by the climbing frame, the second by a giant ladder. The climbing frame is hung with ropes and a hammock. There's a shelter somewhere, according to the sign, and I pick out the stream that bubbles down the middle of the island between rushes and boulders, which goes on to aerate the moat acting as a natural barrier, since gorillas can't swim.

I sneak a look at J.P. and he sits as before, but his head is lowered.

It's a quiet weekday at the zoo. The bright early morning has turned grey, dark clouds rolled across the sky by the wind. The first visitors are trickling in – a troop of mums. They gather with babies in prams and pushchairs, trudge in tired groups and sink gratefully onto the picnic benches clutching cappuccinos and lattes. I catch talk of sleeping regimes and potty-training; much like other species of primate, they bond in social groups, sharing parenting tips and company.

The picnic tables overlook the macaque enclosure, next door to J.P. The same shape and size, but with a way sturdier climbing frame.

The mums that are less tired, or wired on their once-daily coffee, notice me lurking and I watch the suspicion fall across their faces as their kids, those that are old enough to talk that is, all ask 'Mum, why's that man hiding?'

I ignore them. Not sure what to say. No one wants to hear the word *'escape'* when they're sleep deprived, while keeping an eye on little toddlers.

The macaques have amazing colouration too – nothing in comparison to J.P.'s blue bollocks, but they have distinct manes of white-grey fur, flared out around solemn faces. The rest of their coat is a shiny black, tails short and jaunty, with bell-pull ends. They stroll around their enclosure, turning over logs and stones, happy to forage on the floor. The dominant breeding male is big and bulky; he struts around with an air of puffed-up self-importance. All four females are in oestrous, the point in their reproductive cycle when they are most fertile. As a result, their rumps swell up. The alpha male is driven to mate them as much as possible so as not to miss the moment, and so today, on this gloomy morning, he is shagging constantly as the tired mums stop in front of his enclosure. He grabs a female, smacks her tail aside and shrieks as he thrusts away, punching and biting her back, constantly shadowed by his young son, long and rangy, masturbating in support.

'What are those monkeys doing, Mummy?'

'They're just…'

Hysterical kids' laughter. 'That one's pulling its winky.'

'Why is that man hiding?' one of the mums asks.

J.P. sits there on the platform, eyes locked back on mine, and I hear the clump of Bob's boots. 'Hey mate,' he says. 'Got you a coffee.' He holds up a takeaway cup. 'You do drink coffee, don't you?'

'Yeah, thanks,' I say.

He hands me the warm cup and takes a radio off his belt. 'Going to give you this as well.' He offers up the promised two-way radio. 'They're taking Kera out on the island.' He points at the gorilla enclosure. 'We're ninety-nine per cent sure J.P. is getting through the electric fence outside, but if he's slipping out through a loose roof panel we've missed or pushing his way behind the slide somehow, we need to know, so if he does anything odd, radio number 14 and tell them.'

I nod.

'You know how the radio works? Just hold down this button and talk.'

'Okay.' The radio gives me some recognized role at least, gives my lurking a certain legitimacy when the mums come past. 'The macaques are mating,' I say to Bob.

He snorts. 'Rubs his dick raw every month,' he says. 'That's why he keeps biting the females as he's doing it, had to explain it all to this teacher last time they was all in oestrous. Didn't tell her how much he likes it when you put the ointment on for him, though. Easiest animal you'll ever medicate.'

'What?'

'Radio if J.P. does anything dodgy – can't have him on the island when they go out there with Kera.' He closes the hatch to the outside area of J.P.'s enclosure and padlocks it shut.

J.P. stares at the island and eventually we both see the metal doors in the brick gorilla house swing open. Two keepers step outside onto the island and one of them is Hannah from earlier. She has a gorilla on her back.

Kera is four years old, about the size of one of the toddlers tottering past. The size human kids are when they can just about walk, pushing their own pushchairs, the stage where they're not quite steady on their own two feet yet. Kera is that sort of size and weighs 39 kilos. Arms around Hannah's neck, Kera peers over her shoulder. They take her toward the climbing frame.

Later, I will learn that Kera arrived at Bristol Zoo not long before I did, from a hand-rearing facility in Germany. She was born one of twins and their mother rejected them both. Any baby gorilla that needed hand-rearing at that time would be sent to this nursery-type facility and they'd all grow up together and have gorilla company to play with. Kera and her twin brother both grew up there and became too strong and disruptive to stay at the facility, but at four years old they were still very young to introduce into existing gorilla groups elsewhere.

Kera came to the zoo and the process of introducing (or 'mixing') her had begun. Getting her familiar with her surroundings was an essential part of the process, before being mixed with the other animals, so she would know escape routes if she needed them and didn't get herself stuck in a dead end if she was attacked. Or touch the electric fence.

Not that I know any of this right now. All I see are two keepers seemingly playing with a gorilla on the island. What I do know is that I want this life.

Hannah and the other keeper take Kera back inside and pull the door shut behind them.

J.P. is asleep on his platform.

The macaques still shag away next door.

The two-way radio I was given by Bob crackles in my hand. A panicked call. *'Mammal keepers to ISO.'*

I hear the echo of footsteps as Bob sprints into view.

'Copy,' he pants and his keys splash out of his pocket as he passes me at full pelt. I sweep them up and run after him as he charges toward the gorilla house.

The mammal keeper team are gathered at a window. I glimpse Hannah inside, on her hands and knees. Kera, the little gorilla, has her in a headlock.

I hold out Bob's keys for him and he disappears inside the gorilla house.

Hannah throws Kera off and she leaps straight back at her, mouth open in a manic toothy gape. I can see how ludicrously strong she is for something so small; her arms wave, hands clutch and she thrashes with Hannah across the floor. I hear doors clang inside, bananas fly everywhere and Bob bursts in.

The keepers gathered at the window are tense.

Hannah is on her back and shoves the exuberant gorilla away with her foot.

Kera strips off Hannah's wellington boot and tosses it over her shoulder.

Bob scatters a double handful of grapes and walnuts, and Kera makes a grab at them as they bounce across the floor around her.

Hannah abandons her boot and Bob helps her exit the enclosure.

The keeper team all exhale and I feel my heart pound.

The keeper-door shuts with a distant clang and Kera kicks the window we're all staring through, clapping and laughing, picks up the discarded wellington boot and bites it.

Hannah and Bob both step out of the gorilla house and stand in front of the team.

Hannah is slightly lopsided. 'That's the last time any of us go in with Kera,' she says.

I walk home, disappointed. I will have to wait a whole week before my next day in as a volley. My head is full of primates and names.

Chapter 2

Watercolour Challenge 2011

This is a book about Kera as much as it is about hand-rearing and the life I began as a gorilla keeper. I got a job on the mammal team after three months of volunteering. The mammal team covered seven subsections, one of which was gorillas. I worked my way through all the subsections, most of which were species that were 'hands-on', which in zookeeper speak means a species that you can share space (or 'work in') with – ones that aren't potentially lethal. I learned to hide my pen from the troop of thieving squirrel monkeys as they flooded down from the high boughs of the spruce tree on their island to rob the food bucket. I built weighing frames for the aye-aye to sit on; developed a relationship with a gentle old okapi, lifting his velvet-furred stripy legs for daily hoof-care sessions; became obsessed with the colony of naked mole-rats; bottle-fed two lion cubs – but this book is not about any of that. Three years later I started to train as a gorilla keeper.

Not everyone has the right temperament for gorillas, as their society, much like ours, is highly complex. It's essential to have a good understanding of primate social behaviour and most importantly, an automatic sense of safety.

Working in zoos is a profession that carries the risk of death. It's not unusual for zookeepers to be killed by their animals, generally by elephants or big cats, but apes are equally dangerous. An adult gorilla is many times stronger than a human. As a comparison, a muscle-bound man could lift 400 kilos; an adult male gorilla can not only lift objects twice this weight, but also chuck those things around. Telegraph poles, for example.

Before working with gorillas, rechecking locked doors must be second nature, along with an ingrained sense of spatial awareness around the mesh that divides you from the animals you're working with. 'Mesh' in zookeeper speak refers to the metal bars between you and the animals. It comes in different strengths and sizes, depending on the danger of the animals.

Kera was now seven years old, a sub-adult and not yet fully grown. As she had been hand-reared at the nursery facility for those first four years of her life, it meant she had never witnessed adult gorilla social behaviour and as a result she struggled to fit in. She was ostracized by the rest for reasons she couldn't understand and lived in fear of Jock – the silverback – 200 kilos, or 31 stone if that's clearer, a mass of muscle with canines as long as your thumb, capable of cracking open coconuts. Kera's behaviour put Jock on edge; he was prone to attack her horribly for any slight infringement and she spent much of her time alone, plucking out her fur in stress and boredom.

The gorilla house sits in the middle of the zoo. Once home to elephants and giraffe, it was redesigned for the gorilla group and expanded to include the mammal team messroom. We all gather here each morning, upstairs above the gorilla kitchen, and sit at long tables, piled high with section diaries, scattered with empty tins of biscuits and chipped mugs, generally from the gift shops of other zoos. A selection of mismatched chairs, spare wellies, a rack of unused climbing harnesses and a desk for the computer, surrounded by banks of two-way radio chargers. The messroom has windows on three sides that overlook Gorilla Island, seals and the flat felt roof of the hippo house, rippled with standing water, the glass peaked greenhouse of Zona Brazil beyond.

There is a huge monitor mounted on the messroom wall, next to the notice board. The images on it are split across 16 live camera feeds, throughout the gorilla house. I'll be walking about in there in a minute, after the morning meeting wraps up, on my first day as a trainee gorilla keeper. I feel the wave of enthusiasm and excitement. I know I'm about to have the time of my life.

The team all mill about with cups of tea and cram in slices of burned toast. Seal keepers, okapi keepers and the small mammal team.

Bob is leading the morning meeting today, a torn page of scrap paper covered in his scrawled loopy handwriting. 'Right,' he says and the general hubbub falls quiet. He squints at the paper. 'Can't read me own bloody writing.'

We wait.

'Oh yeah, vet's down for hoof care at nine-thirty, so expect them about ten.'

Up on the monitor I see Kera, pirouetting and yanking on a frayed rope.

'Check foot dips.' Bob reads down the list. 'Weigh squirrel monkeys, hippo, armadillo and the tapirs.' He pauses. 'Did the capies get done yesterday?' He glances up. 'No? Okay, weigh those Brillo pads as well. I'll come give you a hand.'

A gorilla face fills one of the screens as it peers into the camera. I can't tell who it is. Komi, maybe? Jock's son. A ridge of wrinkled skin above two dark eyes. Soft fur off a domed head. Two big nostrils above smiling lips.

'Day two-of-three pre-export faecals from the J rats,' Bob continues. 'And the maintenance team are coming down to rod the drains in Twilight. Fuck knows when they'll turn up, break time probably.'

The gibbons whoop in the distance.

'What else we got,' he mutters. 'Stores will need a hand unloading, they'll give us a radio shout when the pallets come in, whole load of Old World monkey pellet, bin bags and a new detergent we're trialling.' He nods to himself. 'Don't use any today, though, I've got to work out the ratios.'

'Have we got a rough time for that?' Hannah asks.

'Lunchtime probably,' Bob says and looks back at his list. 'Day two-of-three pre-export... no wait, I've done that one. Ten forty-five there's a student needs meeting at Monkey Jungle, wants to talk about those idle bloody howler monkeys, so good luck with that, and we've got ladder training this afternoon, three of us got to go at least.'

Moans and groans meet this comment. The only ladder I'll be climbing up today is the giant gorilla one from the floor of den two. Jock leans against the rungs right now, I can see him up on the screen.

'Vet visuals this afternoon all up at Twilight World,' Bob continues. 'Spot-on treatment for Righty and Lefty, med review on Mr Ben and get them to check out Rasputin's thumb while they're there and that's it.'

Chairs scrape as everyone stands up and queues to pluck two-way radios out of the chargers. A series of bleeps as the team switch them on.

'Oh,' Bob calls. 'Them J-rat faecals are in-house parasitology, the form's on the pin board.'

The team tramp downstairs and head off to their separate subsections around the zoo.

I wait with Bob and Hannah. I'm excited, the first step to becoming a gorilla keeper. Fully trusted.

'We've got this picture thing to do today,' Hannah says.

Bob rolls his eyes and slugs back the dregs of his coffee.

'Picture?' I ask.

'Animal drawings,' Bob replies. 'Zoos are selling them for a fortune apparently.'

'What?'

Bob ditches his cup in the sink around the corner. 'Drawings done by animals.'

'The marketing team,' Hannah says, 'have asked us if Sal will paint a picture.'

'With what?'

'Watercolours,' Hannah explains. 'Sal's really interested, we've painted some on the floor in the keeper-corridor and she's been watching, so we're hoping she might do one of her own.'

'A picture?'

'I thought those were the ones Sal did.' Bob winks at me.

Hannah raises her eyebrows. 'Art's not my thing.'

'Sal looks like van Gogh,' Bob says, 'missing half an ear, toothless old bag.'

'Bob,' Hannah screeches. 'Sal's lush.'

'What's she going to paint?' He laughs. 'Gurt big cake? Bunch of bananas, bucket of fried chicken?'

'Ignore him,' Hannah says. 'We're going to try and get her to do it now, before we shut them all out for breakfast.'

'I'll leave you to it for a bit then.' Bob heads for the stairs and stops to turn round and look at me. 'Be careful in there, mate,' he adds. 'Don't lose any fingers.' He clatters off down the stairs and we follow him into the gorilla kitchen.

Hannah locks the door behind him. 'I'll treat you like you don't know anything,' she says. 'Might sound patronizing, but I'll know I haven't missed anything out, if that makes sense?'

'Absolutely,' I agree, notebook in hand. Neither Hannah nor Bob ever make a big deal of being classed as senior keepers; the very reason we all listen to them so intently.

'I've done you this.' She reaches up to the top metal shelf, stacked with plastic tubs of meds and pregnancy kits. She passes me a sheet of paper, a plan of the gorilla house ground floor with neat boxes and arrows. 'So you can get a handle on all the different dens.'

Gorilla house, ground floor

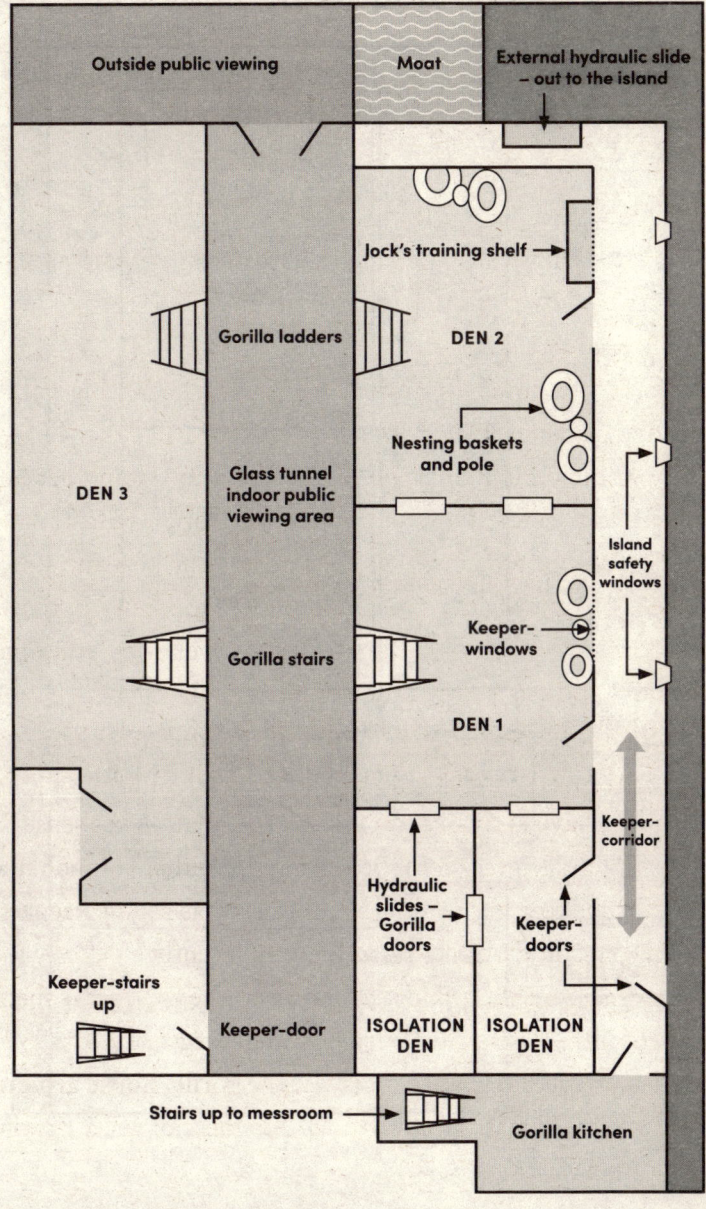

Outside public viewing

Moat

External hydraulic slide – out to the island

Jock's training shelf

Gorilla ladders

DEN 2

Nesting baskets and pole

Glass tunnel indoor public viewing area

DEN 3

Island safety windows

Keeper-windows

Gorilla stairs

DEN 1

Hydraulic slides – Gorilla doors

Keeper-corridor

Keeper-doors

Keeper-stairs up

Keeper-door

ISOLATION DEN

ISOLATION DEN

Stairs up to messroom

Gorilla kitchen

Gorilla house, first floor

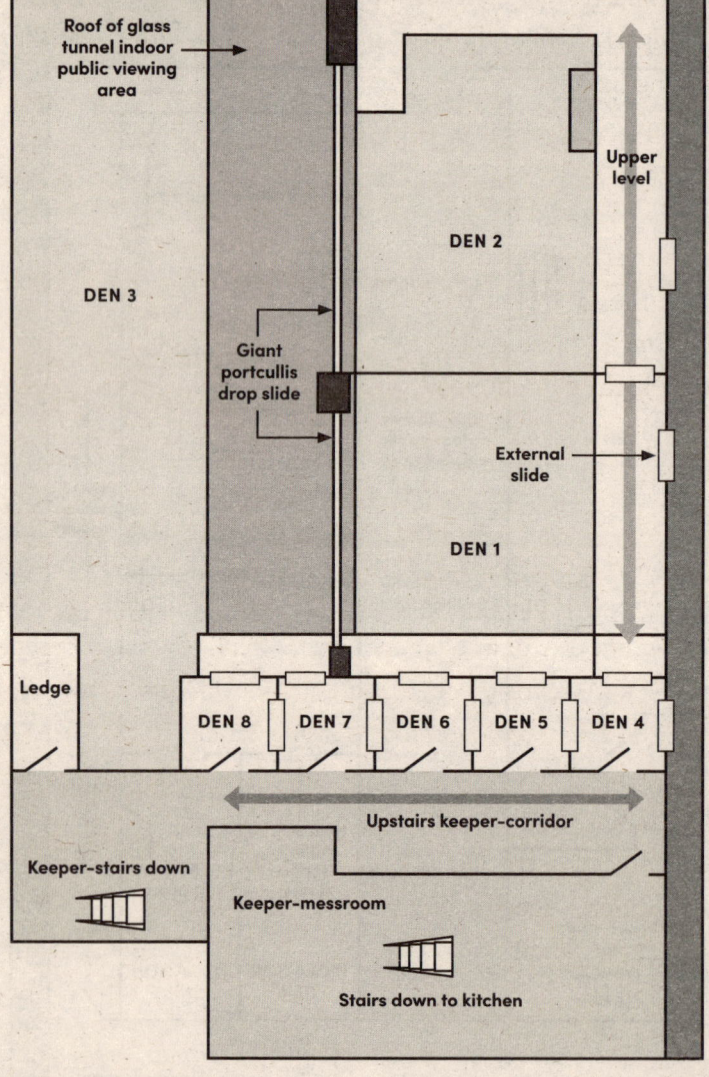

Roof of glass tunnel indoor public viewing area

DEN 3

Upper level

DEN 2

Giant portcullis drop slide

External slide

DEN 1

Ledge

DEN 8 DEN 7 DEN 6 DEN 5 DEN 4

Upstairs keeper-corridor

Keeper-stairs down

Keeper-messroom

Stairs down to kitchen

'Dens' in zookeeper speak refers to internal enclosures, like rooms in a house. I look down at Hannah's plan and my mind reels; I concentrate on the ground floor and see a confusing array of different dens and slides, with ladders and nesting baskets.

'Staple it into your notebook later,' Hannah says. 'First things first. Breakfast.' She leans against the deep sink. 'Bob's been and chopped this all up for us already, but most mornings it's the first thing you do.'

On the shelves next to the sink are two big plastic crates, with metal handles at either end, so they can stack. One is piled high with chopped vegetables; carrots, parsnips, leeks, cucumbers and peppers. The second crate is full of halved iceberg lettuces and bunches of coriander.

'We'll scatter a third of their pellets out too and keep the rest back for later.' She crouches and holds up one of the 'Old World monkey pellets', the primate equivalent of a dog biscuit, the size of a sugar snap pea. 'You know the difference between Old World pellet, and New World?' she asks.

'Sure.' Old World monkeys come from either Africa or Asia, whereas New World monkeys live in South America. The pellets we dish out to them all as part of their daily diet have precise nutritional requirements and New World species have different needs. 'Do they like their pellets?'

'Yeah,' Hannah says. 'Bob likes it too, he's always eating handfuls.'

'He eats everything.' I fold up the plan and see the upper floor is printed on the reverse of the page, with even more slides and arrows.

'We'll lock them all in first,' Hannah reassures me, 'and do the perimeter check out on the island.'

I nod and write in my notebook.

'Then we can scatter their breakfast across the island and they'll all toddle outside. We lock them out and clean. If it's tipping it down, or in winter when it's freezing, there's a whole other routine, but I'll teach you that later.'

'Okay,' I say and add *Dry weather gorilla routine* as a title to what feels like is going to be pages of upcoming notes.

Hannah steps past me to the metal safety door, which leads into the downstairs keeper-corridor. It sits within a steel frame and has a small eye hatch.

'Have a look through this first. Make sure the gate inside is shut and there's not a gorilla sat waiting for you.' She holds open the spy hatch and I peer into the gloomy vestibule. A monitor screen hangs on the wall, rakes and a net on a thin metal handle lean in the corner. A reinforced mesh door inside blocks access to the keeper-corridor beyond and is padlocked shut.

'Happy that it's locked?'

'Yep,' I say.

Hannah closes the hatch, unlocks the metal door and pushes it open.

A heavy scent of cut grass, shit and urine rolls over us.

Hannah stands in front of the wall-mounted monitor. 'We've got three of these,' she tells me. 'The same as the one upstairs in the messroom.'

'Okay.' The screen is split. Four by four.

'It's confusing as hell,' Hannah says, 'but you'll get it eventually. The main thing is, you can see where everyone is before you open this next inner gate into the keeper-corridor.'

I pull the plan out of my pocket and unfold it.

'There are blind spots everywhere,' Hannah continues and slows down so I can comprehend what she's imparting. 'The upstairs level runs directly above the keeper-corridor.' She points up at the pocked concrete ceiling. 'On the other side of the page,' she adds.

I spin the page back over and see the text box with *upper level* and more arrows.

'You've always got to think safety first with this lot,' Hannah says. 'If they get hold of you, you're not coming out in one piece.' She points at the monitor. 'I've got Jock here, look, in den two.' She points at the huge sleeping pile of gorilla on the floor by the giant ladder. 'Kera is usually in den three in the mornings, with the zoomies, but I can't see her.'

'Zoomies?'

'When they zoom about,' Hannah explains. 'All wound up and excited. Ready for breakfast.'

I hear a succession of grunts from the other side of the safety door. A gorilla grumble. A repeated 'urm, urm' sound.

'Alright, Sal,' Hannah calls. 'Be with you in a minute.'

'Is that Kera?' I point at the top left feed.

'That's Cheery Komi,' she says. 'And Moki's down here.'

'How can you tell?' Jock's two kids look identical to me.

Hannah smiles. 'You'll recognize them soon. Cheery Komi is Jock's son, he's a bit rangier, long and skinny, compared to Moki. She always looks grumpy.'

I look back at the screen. Jock's two kids are about the size of a labrador to give you some comparison.

'It's about behaviour as much as anything else. Komi is a happy-go-lucky kind of gorilla. Friendly, you'll love him. That's why he's Cheery.'

'And Moki?' I start a new page of my notebook. I write *'IDs: Jock enormous'* and *'Komi cheery.'*

'There's a phase the sub-adult females go through,' Hannah elaborates. 'Moki is Jock's favourite, his first child and so she never gets told off. Romina, her mum…' Hannah pauses to scan the multiple camera feeds. 'Here's Romina, look.' Hannah points and I make out an adult female gorilla asleep on its back in a nest of wood-wool, twice the size of the kids. 'She's the kindest gorilla you'll ever meet, but she never told Moki off either.'

I write *'Moki'* in my notebook.

'It means Moki instigates most of the fights,' Hannah continues. 'You won't love her quite so much.'

'Do you get many fights?'

'Jock's going through a phase of hating Kera,' Hannah says. 'You know she was hand-reared until she was four?'

I nod.

'Moki does all she can to get Kera into trouble, because she knows Jock will go ballistic.'

I add *'troublemaker'* to Moki in my notebook. 'Bob says working gorillas is all about managing Jock's mood?'

'It is,' Hannah agrees. 'And managing everyone else's to a degree, all to take the pressure off Kera.' She points at the monitor again, her fingernail chipped. 'Kera's over here. She's smaller than Sal or Romina, but bigger than Cheery Komi and Moki, and she's got a square head.'

I write *'Kera square head'* and then *'Romina kind'*. 'That's the head count done, then?' I say.

Hannah nods. 'We've seen them all and we know Sal is in the den just through the next door, so we're good to open up.'

I add *Sal* to my book. Bob always goes on about her being obese, but I wait to meet her before writing anything else.

'This,' Hannah says and plucks a long remote control from a wooden holder on the wall, 'will be the bane of your life.'

The remote is three times the size of a TV control.

'Each button does two separate things. Opens or closes one of the doors inside the enclosure. Slides, we call them.' She pauses. 'Write that down. Slides are the hydraulic mesh doors inside the enclosure.'

I check the plan again briefly. There are loads of slides and some with arrows that point outside. I scratch away in my notebook.

'Sliding doors,' she repeats and holds up the remote control. 'Each one operated by this.' There are 20 buttons, in two columns of ten, all numbered and lettered in different colours.

'This button at the top turns the system on and this second button here switches between suites.'

I've now got no idea what she's on about. 'How the hell do you remember that?'

She smiles. 'You'll get the hang of it.' She pauses and holds my gaze. 'The key thing to always remember, without fail, is to see the slide you're opening or closing.'

'Okay.' I underline *'watch slides'*.

'You don't want to squash anyone, trap hands or feet.'

She unlocks the safety door and leads me into a long brick corridor, which runs the length of the house, with a red fire hose on a reel halfway down and tools hanging from the wall. In zookeeper speak, 'keeper areas', or in this case 'keeper-corridor', are mainly out of sight to the public and have doors to enter the enclosure, to clean. Each of the gorilla 'keeper-doors' sits within a steel frame, made of mesh, with a locking arm and two dangling padlocks. The mesh throughout the gorilla house is of a gauge that stops an adult gorilla getting its full hand out, but allows them to reach through with their fingers.

Hannah stops at the first keeper-door. Sal leans against the mesh, drinking from one of her own long boobs. Rolls of fat surround her neck and belly. She grunts at Hannah and shifts her huge bulk to stare at me, her mouth still clamped around her nipple.

'Don't be disgusting,' Hannah says and adds a perfect imitation gorilla grunt back at Sal in reply. 'She can be funny sometimes,' Hannah warns me. 'Gets possessive over her keepers, so watch out she doesn't try and grab you through the mesh.'

Sal stands up on all fours. Her knuckles are a ridge of dark grey skin, she is barrel-shaped with a long conical head. She waddles on bandy legs out through a low doorway in the concrete wall between dens.

'Good girl, Sal,' Hannah says and holds up the remote control for me to see. Her fingers press buttons with practised ease and a mesh door slides shut after the departing Sal. 'We're trying to get weight off Sal, but I'll fill you in on all that later.'

'Okay,' I say.

'This is the isolation den,' she continues. 'It's marked up on your plan. Sal usually waits for us in there.'

A second slide at the back of the isolation den shudders closed with a further hum of hydraulics.

'All the dens have got more than one entrance,' Hannah explains. 'You don't want to create any dead ends in here, just in case it all kicks off and they have a fight.'

I turn the page of my notebook.

'I always go and say hello to Jock first,' Hannah says, 'before I do anything else, get a look at him and see what sort of mood he's in, let him know I'm here.' She stops at a mesh window which looks into the enclosure. 'Keeper-windows.' Hannah waits for me to write in my notebook.

I peer through the mesh window. Sal squats for a piss. The floor inside is scattered with torn paper sacks and stringy clumps of bedding. There is shit everywhere, dried trails of diarrhoea down the glass public windows. Broken sticks and the odd half-chewed turnip.

'They can get their fingers through this mesh.' Hannah stands at the window. 'Not a whole hand, well the younger ones can, but the gauge is too small for an adult gorilla to get a hand out and grab you. But—' she pauses— 'you still have to be super careful no one tries to hook you with their fingers.'

I write 'MESH' in my notebook and underline it.

'This is den one,' she says. 'Next door is den two and over across the hide is den three and then we've got a row of feeding dens upstairs.'

I look back at my plan.

'The hide?' I say.

'The glass over the top of the public area. We call it the hide for some reason. I guess like a bird hide for the public. I'll let Bob tell you how this used to be an elephant and giraffe house.'

Bob loves to go on about the history of the zoo, how the enclosures have evolved over the years as zoo husbandry has changed.

'That's why it's so great for the gorillas,' Hannah says. 'They love all the height.'

Sal clambers into a thick steel nesting basket on her side of the keeper-window. The baskets are giant bowls of curved smooth bars, attached to a huge steel pole, with hand and foot pegs to help the gorillas climb them. The poles all reach to the upper level of the enclosure, with a second identical high nesting basket. In the wild gorillas make nests in trees to sleep in, secure and safe above the forest floor. There are six of these huge steel baskets throughout the gorilla house, giving the animals a choice of nesting sites.

Sal leans forward in the basket and slides three smooth grey fingers through the mesh. She frowns at me with dark brown eyes, deep beneath her curved eyebrow ridge. The fur on top of her head is sparse and brushed with dandruff.

'While we're on windows,' Hannah adds, 'you've got these safety windows behind you.' She points down the far wall of the keeper-corridor at a row of letterbox windows, each with a deep sill in the thick concrete. 'They give you a view out onto the island.'

I stoop to peer through one as Hannah leads me down the keeper-corridor to a second mesh window.

Jock strides across the floor in den two toward us, swift and intent. He's enormous. I've seen him through the glass from the public area of course and out on Gorilla Island, but his sudden, advancing proximity makes my toes tingle. Massive shoulders, and dark fur down his woolly arms. Head twice the size of ours. He stops as he treads in a pile of shit, carefully wipes his foot clean with a handful of wood-wool and glowers at some gorillas out of sight, above our heads on the upper level.

'Good lad,' Hannah says. 'Al's come to meet you.'

Jock pulls himself up onto the training shelf, a table of oak beneath the keeper-window. He fills the space. His thighs are covered in silver fur, his chest a bank of muscle. He looks down at us both through the mesh window. His eyes dismiss me and he focuses on Hannah.

'What have we got?' She reaches into her pocket.

He holds out his palm on the other side of the mesh, a huge dish of sausage fingers and notched black nails.

Hannah deposits a handful of pellets through the mesh.

'Do they try and hook you with their fingers?'

'Not usually,' she says.

Jock licks the pellets off his hand and cups his palm for another load. His eyes dart up as feet shuffle overhead and Cheery Komi pokes his head over the ledge above. Even upside down his face has a smile.

Jock's huge chest vibrates as he coughs at his son. A repeated low grunt.

'That noise means piss off,' Hannah says. 'It's a warning.' She piles a second handful of pellets through the mesh and leads me back up the keeper-corridor. 'Right, let's see if Sal will paint us a picture.'

Sal has reclined in the nesting basket, a collar of flab across her shoulders, arms resting on her big sack of belly. Her dark, beady eyes fix on me.

Hannah gorilla grumbles at her, the repeated 'urm, urm' noise. 'We're going to get the paper and stuff into the isolation den,' she says.

I look through the mesh keeper-door. 'Kera had you in a headlock in there on my first day.'

Hannah laughs. 'Was that your first day?'

'I was trying to figure how J.P. kept getting onto Gorilla Island. That was my first job here.'

'Bob was right, wasn't he?'

I nod. 'He was slipping between the strands of the electric fence outside.'

'Bob loves to go on about being right,' Hannah mutters.

Sal pounds her fist into the mesh next to my face.

A sudden crash that makes me freeze.

'Sal,' Hannah chides. 'Al's nice.' She looks at me. 'You took that well.'

I fizz from the sudden rush of adrenaline. 'Like a sledgehammer.'

'They're so dangerous,' she says, 'but they'll get to know you.'

I try my first gorilla grumble, growl urms.

Hannah raises an eyebrow. 'I'm checking no gorillas are in here,' she says with a nod into the isolation den. 'Nobody hiding around the corner and both the slides are shut.'

She points at the two mesh gorilla-sized doors linking the isolation den to the rest of the enclosure. 'We're safe to go in.' She unlocks the keeper-door. 'Can you grab the paper and paint? It's in the kitchen on top of the fridge.'

'Sure.'

'Get two sheets of paper and all the brushes, I'm going to paint one at the same time out here in the keeper-corridor.'

'No probs,' I say and head back out into the kitchen, the impact of Sal's fist still ringing in my ears. A piston of raw power. If she punched me, my head would come off.

I open my notebook again.

Jock enormous.

Komi cheery.

Moki troublemaker.

Kera square head.

Romina kind.

Sal dangerous, I add.

'Get some paint too,' Hannah calls. 'We've been mixing it up in old yoghurt pots.'

Hannah has swept a clean patch on the floor in the isolation den. I hand her the paper, an A3 sheet. I've mixed up some green, blue and black paint, in three separate pots.

Sal peers in at us through the mesh slide.

'What's this, Sal?' Hannah says and lays the paper on the floor. Places a brush on top and lines up the pots of paint.

Sal grunts at Hannah and her tone sounds interested, friendly even.

Another gorilla face appears over her shoulder. It's Cheery Komi again, come to have a look. He stands up on his long, thin legs, scrawny and on tiptoe, to take a look at what's going on over his mum's shoulder.

Hannah steps back out and closes the keeper-door, clips both padlocks closed and tugs on them to make sure they're shut. 'We don't want you, Komi,' she says. 'Your mum's the artist.' She lays the second sheet of paper on the floor, just outside in the corridor, and gets her own paints set up.

Sal coughs at her son, the same tone Jock used earlier, and Cheery Komi scurries out of arm's reach, a hopeful smile on his face. He hovers by the steps that lead up on top of the glass ceiling over the public area.

What had Hannah called it?

The hide?

The gorilla house has a broad tunnel of glass for the public to walk through, den one and two on one side, with the long den three opposite. The hide has windows that look into the dens on either side and thickened glass panes above the visitors' heads. They can look up to see gorillas strolling overhead. Ooh and arrgh at the sight of Jock's enormous knuckles, whoop and groan when one of them takes a piss.

Hannah mirrors the layout prepared for Sal and sits cross-legged by the keeper-door. Her fingers dart across the long remote control and the slide in front of Sal grinds open.

Sal pushes her way in and Hannah shuts the slide behind her, so Cheery Komi can't slip in and disrupt everything. Hannah hands me the remote. 'Don't press any buttons,' she says and picks up the paint brush. Holds it up so Sal can see it. 'What's this, Sal?'

Sal slouches forward on her elbows to sniff the sheet of paper and glances through the mesh at Hannah.

'Like this.' Hannah dunks her brush in a yoghurt pot.

Sal picks the sheet of paper up to peer underneath it.

Hannah dabs a green line with her paint brush. 'Like this, darling, look.'

Sal tears the paper in half and crumples it into balls.

'Wait,' Hannah says as Sal jams the first ball of paper into her mouth, 'don't eat it.'

Her jaws work on the wodge of paper and she picks up a yoghurt pot of paint, slugs it back like she's out drinking shots and grins at us. Her few remaining yellow teeth are stained green.

'Wait, Sal, no.'

Sal knocks back the next two pots of paint, downs them one after another, jams in the second ball of torn paper and begins to chew it to a pulp.

'Sal.' Hannah uses her chiding tone again.

She continues to chomp and gives Hannah the side eye.

'Will she be alright eating all that paper?' I ask.

'Sal, no.' Hannah's tone has gone sharp and precise.

The long, grey fingers pick up the paint brush. She bites the end off, chews her way through the wooden handle and discards the metal collar and bristles.

'Shit,' Hannah mutters.

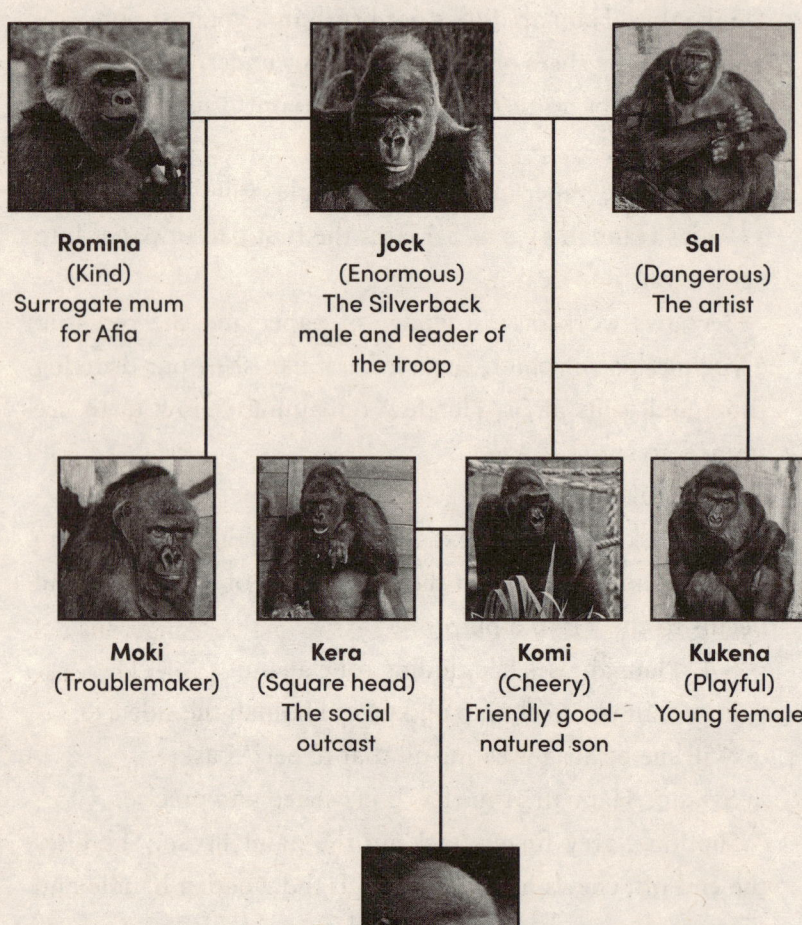

Romina
(Kind)
Surrogate mum
for Afia

Jock
(Enormous)
The Silverback
male and leader of
the troop

Sal
(Dangerous)
The artist

Moki
(Troublemaker)

Kera
(Square head)
The social
outcast

Komi
(Cheery)
Friendly good-
natured son

Kukena
(Playful)
Young female

Afia
Born by C-section

Chapter 3

A Day with the Gorillas

I have been working with the group for three years when we first notice a change in behaviour between Kera and Jock's son, Cheery Komi. Kera is now ten years old and fully grown. The hair on her arms has begun to grow back, she's stopped plucking out her own fur because she is spending an increasing amount of time with Cheery Komi, who is aged eight and retains his good nature. His mum, Dangerous Sal, gave birth soon after I'd started training as a gorilla keeper to her daughter Kukena, who is at the playful-exploration phase, happy, confident and full of energy, but tiny in comparison to everyone else.

The zoo clock chimes eight in the morning as I open the door into the keeper-corridor. The heavy gorilla aroma engulfs me for another day and I see myself step inside on the monitor screen. It's developed a slight time delay, flickers to catch up through the 16 different camera feeds in the usual grid on the screen.

I work my way methodically through the feeds, doing the headcount before unlocking the mesh door into the keeper-corridor. Hannah's plan of the gorilla house is long gone, but she created a new laminated version which we keep on a clipboard next to the monitor.

The crisscross of webbing and ropes in den two bounce on the split screen as Kera clambers high above the enclosure floor. She wobbles and sways, toes and fingers gripping multiple ropes at once as she crosses den two, avoiding Jock below.

Moki, Jock and Romina's daughter and troublemaker-in-chief, clambers into one of the high gorilla baskets, the nest that Kera has just vacated, and adds a pilfered pile of tangled wood-wool to sit on. I recognize her by the arrogant swagger.

Kera comes into view on a different feed. She's on the upper level of den one, above the keeper-corridor, sitting by the outside slide and glancing down over the edge, alerted by the jingle of my keys.

I can smell Sal, waiting for me in the isolation den.

I undo both padlocks and swing the door open into the keeper-corridor.

A broom has fallen off the tool rack.

Sal leans her layers of bulk against the mesh keeper-door as usual and holds an empty yoghurt pot in her long, grey hand, fingernails jagged and clogged with gorilla shit. Her piercing dark eyes gleam out from beneath her sloped forehead, crossed with wrinkles, her earlobes torn and scarred. She grins her couple of long, yellow teeth at me and waits. Expectant. Holds up the yoghurt pot. Willing me to understand what she wants.

I like to start the day with Sal in a happy mood, even though I know she's trained me to dance to her tune. I top up her yoghurt pot with fruit tea and she instantly downs it, like she did with the paint on my first day. One swift slug and it's gone. She holds my gaze and lifts up the yoghurt pot for a refill, necks that one too and keeps knocking them back until she's had the entire bottle.

I grumble at her. My grumbles sound right now, practised and understood.

She grunts in return, stands up on her thin legs and waddles out of the isolation den through the slide into den one.

I pick up the fallen broom and hang it back on the wall next to the floor-squeegees, the wall mops and plastic hand-shovels.

A series of grumbles follow from the others as I stop at the first mesh keeper-window, looking into den one. I see Cheery Komi stood near the public windows, up on his back legs.

He's no longer the scrawny little four-year-old I started working with, he has bulked out and is generally sweaty. Over the years we've developed a private code. He gives me a furtive nod toward the far keeper-window, next door in den two, all the way down the end of the corridor. He chooses his moment, when no one else is looking, the same look every day. Stands up like a person and darts his eyes from mine to the window and back again, desperate not to alert anybody else, particularly his vile sister Moki. *'I'm ready for my drink, see you at the keeper-window – don't tell anyone.'* That was the message he conveyed, our shared understanding of that look.

'Not for a while yet, mate,' I say. 'Got your dad to do first.'

Sal swings her massive barrel stomach into the metal nesting basket in front of the keeper-window, juts her scarred conical head forward and holds up the yoghurt pot again. Gives me her toothy look.

'Where's your kid?' I ask her and peer into the den for Kukena, Sal's young daughter. She sits in a chewed cardboard box and slides it across the wood-wool-strewn floor like she's in a rowing boat.

Sal taps the yoghurt pot against the mesh.

Jock's vast bulk blocks out the light as he ducks through the slide from den two. The pungent cut-grass smell rolls off him. He's puffed up, lips pulled tight, shoulders bulged, and the silver fur gleams down the concave arch of his back.

I know straight away, by the way he stands, that he's on one.

I hear Kera scamper away upstairs above my head and dart next door. Cheery Komi gives me the nod again and also exits as Jock barges in and I glimpse Kukena abandoning her carboard box to scurry off under the set of giant concrete gorilla stairs.

Sal, trapped in the metal basket in front of me, grumbles the reassuring *'everything is fine'* rumble.

Jock bowls past her, thumps her back en route, a deep echoing thud, and hurls the half-torn cardboard box out of his way.

Sal gives me a venomous look, the yoghurt pot crushed in her crooked fingers.

I glance at the oestrous chart, hung on the wall, a day-by-day yearly vinyl calendar that we fill in daily with squidgy pens. The chart plots and records the female gorillas' oestrous cycles, enabling us to discover potential pregnancies and predict Jock's behaviour. I can see that Sal came into oestrous again yesterday, the first time she's cycled since giving birth to Kukena. Unlike the macaques, back on my first day as a volunteer, or female chimpanzees, there isn't a glaringly obvious physical oestrous signal to the males. With gorillas it was way more behavioural.

Jock resumes his puffed-up stance and starts to make his annoying, repetitive raspberry noise, blowing through pursed lips over and over again. We all know how this is going to go.

I realize Kind Romina has been sat by the gorilla steps all along, quiet and out of sight, the crest of fur on the back of her head arrayed in a perfect fanned hairdo, a sudden back-lit halo as the interpretation signs out in the public area blink on.

She pats her belly softly and pulls her lips back. She wants to get past Jock too, join the exodus out of his way, but he blocks both slides and his raspberrying goes up a notch, higher pitched, eyes fixed on Sal.

'Jock,' I call, holding up a second bottle of fruit tea for him to see.

His eyes flick across mine, but he's fighting. The raspberry builds in tone, ascends to an even higher pitch and I see Sal wince.

I grumble.

Sal joins in.

Romina grunts hopefully and pulls Kukena toward her, reaches out for Sal's kid and lets her clutch onto her belly.

Jock can't contain it. The raspberry turns into a bellowing roar. Two beats on the massive plates of his chest muscle and he charges across the den and knocks Sal out of the basket.

She coughs and bares her yellow fangs as she tumbles to the floor but remains submissive as Jock stands over her.

Romina shuffles off behind Jock's back, pissing herself as she goes, and disappears through the slide next door, taking Sal's kid with her.

Jock's thick, damp smell pumps off him and sweat runs out of his eyebrow ridge.

'Good lad.' I wave the bottle.

Sal's jagged fingernails claw at my arm.

She's sidled away from Jock and has her wonky fingers jammed as far as they'll reach through the mesh of the keeper-window, a sly hungry look in her eyes, fixated on the bottle in my hand.

If Sal and I were both gorillas, she would be cleverer than me, likewise if we were both people. She's been around, seen it all, knows how life works, no messing about. Her efforts at communication with us are more pronounced than the others and she uses a form of practised mind control. She's trained me to give her shots of fruit tea each morning. Her efforts are always ongoing. She follows a logical path of behaviour that should, eventually, achieve some glorious breakthrough with us, if only we weren't so bloody stupid.

It's not that she is asking a lot from us, a basic concept that try as she might, we fail to consistently grasp. Sure, sometimes we get it, give her things, her breakfast, her dinner, but Sal finds the way we fail to give her whatever she points at immediately deeply irritating, and when we repeatedly fail, deliberately misunderstand her, she becomes angry, vindicative and holds a grudge.

I need to be extra careful of her today.

Three low coughs sound from den two next door, a warning, and I know it's Kera from the pitch.

I can't see what's going on next door from here, still holding Jock's bottle, keeping away from Sal's filthy fingernails. I don't have to see what's going on to know that it's Jock's daughter Moki causing trouble.

Moki has three main pastimes: sitting about frowning with her arms crossed, because nobody will play with her, winding up her dad and catching wild ducklings from the moat outside. She's trained us to give her tomatoes, at an ever-increasing rate of extortion, in exchange for the safe return of each pitiful, cheeping, quivering handful of downy fluff that she eventually unclasps from her long, grey palm.

In Jock's eyes, as Moki is his eldest child, she can do no wrong, and the rest of the group can't stand her. Even her mum, Kind Romina, spends most of the time acting as if she doesn't know her anymore, playing with Sal's kid Kukena instead.

When Kukena was little, at the wobbly just-beginning-to-explore phase, Moki would lurk, hunched with a grim look on her grumpy face. Kukena would totter away from her mum, reaching for flower stalks, immersed in wonder at the new textures and smells. That's when Moki would strike. Grab Kukena and career off around the island with her clinging onto her belly. Sal would give chase and Moki would deliberately lead her toward an enraged Jock.

Incensed by the wails of terror, Jock would batter Sal in response, deeming her responsible for losing Kukena, because of course it couldn't possibly be Moki's fault.

Jock has also heard Kera's low warning cough from the den next door.

His huge head snaps to look at the slide into den two. In profile his head is longer than it is high, the jaw muscles bulge. He glares at the slide, blood vessels a taut web across his chest.

Much as I don't have to see Moki to know what havoc she's about to cause, neither does she have to see Jock to get her desired reaction.

There's the shuffle of gorilla feet next door and the coughs escalate into ear-piercing shrieks.

Sal jumps at the window and pounds the mesh between us with her hammer fist.

Jock bellows and charges next door, clouts Sal on the back again as he passes. The bottle of fruit tea bounces down the corridor as I grab the tub of pellets and rattle it.

The monitor screen on the wall shows a tumbling ball of fighting gorillas on the upper level above my head and Jock hurtling up the nesting baskets, intent on attack, his massive bulk horribly quick as he launches for Kera. She shrieks and stress shit squirts from every gorilla.

Jock reaches the upper level and grabs Kera, smothers her with his huge body and pins her down, to bite her back.

Kera wails.

And the females all turn in unison.

At the sound of Kera's wail, they all halt and Sal leads the counter charge.

Romina follows, hairdo still perfectly aligned, teeth bared, screaming, Kukena clinging to her belly. Moki joins in too, even Cheery Komi, who was still lurking at the far keeper-window patiently waiting for me to get to him on the fruit tea run. They snarl and swipe at Jock, scream and chase him off.

On the monitor I see Jock abandon Kera and he flees, crashes his way through the slide above my head back into den one and tumbles down into the basket Sal sits in at the keeper-window. His giant chest heaves. The females shriek and gather above his head and I see Romina's face stretched wide in rage.

Jock cowers.

They cough a bit more, someone grumbles and they all wander back next door.

A silverback's behaviour is regulated by the rest of the group. The females refuse to tolerate overly aggressive behaviour. In the wild a silverback has to be able to protect his family from rival males, but his behaviour remains under constant review. If he's an arsehole, he's deserted, replaced by a calmer rival, and they are all, eventually, traded in for a younger model when they get old and dilapidated.

I pull the lid off the tub of pellets and offer him a handful.

He puffs and pants.

'Why did you do it, then?' I ask him. 'You know Moki's trying to wind things up.'

He shifts his huge bulk in the basket, leans toward the mesh, and pushes his soft lower lip out, so I can deposit another handful of monkey pellets.

I take a quick glance at the monitor, but no sign of Kera, or whether she's limping or not.

He pokes me with a giant sausage finger. I've got no fruit tea to offer him as I lost it in the ruckus.

It's 8.09 a.m.

The first rule of gorilla cleaning: don't go in with them. Make sure they're locked outside on the island.

I stand on the podium overlooking the moat and count off the gorillas to make sure none of them have stayed hidden inside.

Kind Romina, despite clutching a lettuce and a carrot with her toes, still moves in her stately way across the sandpit at the front of the island, picking up pieces of cucumber and parsnip. I've cut all their breakfast super small this morning and scattered it across the island, to keep them all content and busy while I clean.

Kukena has returned to her mum and rides on her back, trailing an arm as Dangerous Sal strolls through the pampas grass. Jock wanders along the ridge, plucking up bits and pieces as he goes, calm and slow, the fight forgotten. Kera joins Cheery Komi on top of the climbing frame, clambers up with an armload of food, and I can see both her legs are moving normally. She sits next to him, the two of them side by side, and there are no visible injuries, but I'll get a close look later.

Moki, the troublemaker, is at the base of the huge tree and keeps tabs on everyone as she stuffs her face with yellow pepper, her long hands protecting the food she's gathered.

No matter how much you dress it up, a certain proportion of a zookeeper's day is spent picking up faeces of one sort or another. Zookeepers refer to excrement as shit to each other, poo if asked a question by the public and faeces if you're talking to the vets.

Everything is swept from the upper levels onto the floor; old bedding, torn cardboard boxes and branches, or 'browse' in zookeeper speak. Browse is part of the gorillas' daily diet, bunches of freshly chopped leafy branches, stripped of bark by the time they've finished with them. Everything is swept into piles and taken outside bag by bag to be dumped later.

Soaked in sweat already, I stoop to take a look out of the letterbox window in the keeper-corridor. The narrow view catches Moki, still leant against the battered tree, elbows on her knees, glowering as usual.

The tiny window is suddenly filled with Kera's square face and she slaps a flat hand against the reinforced glass.

Cheery Komi peers in at me over her shoulder and the two of them mooch off together across the front of the island. They look suspicious, behaviour muted, like they're trying to keep a low profile. No sign of bites to Kera's shoulders either by the look of it, but I need to check her back.

Cleaning the huge steel nesting baskets is next on the list. Each giant bowl of bars drips with stress shit. 'Thanks, Moki,' I mutter. I talk to myself all the time – not if the gorillas are inside, then I talk to them, but I've got some weird thing about silence, we all do. We're better off in groups, like any social primate. Nesting baskets blasted with the firehose, washed and rinsed, and it's back out into the long keeper-corridor to grab fresh wood-wool.

'Wood-wool' comes in slabs, packed into a bale, bound with electric-blue plastic string. Extra-fine shredded wood shavings, wool-like, and it smells great, a tingle of resin to momentarily displace the heavy stench of shit and urine. It's stored at the far end of the keeper-corridor and I stop to glance out of the window, a peek across the front of the island and the sweep of rushes along the water's edge.

Moki is standing in the rushes, troubling the ducks. She reaches a hand into the water and a babble of frantic quacks erupt that I can hear through the thick, heavy gorilla slide that gives access out onto the island. A duck leads her brood out onto the moat and Moki stands up on tiptoe, both forearms wet to the elbow, and peers after the retreating line of ducklings.

Still, she's not causing havoc for Kera or any of the rest of the gorillas, and I get on with the rest of the clean; hose, scrub and rinse. Throw in today's browse quota, scattered throughout the entire house to avoid any flash points. Check all the keeper-doors are locked and then I slow myself down and check them all again before turning on the giant remote control. I open up the external slides onto the island. The gorillas bustle in to investigate the browse and we all grumble as we move about. I still need to check Kera properly and scan across the camera feeds to see if she's on the upper level. The monitor drips from my over-enthusiastic hosing. No sign. I rub condensation off the window onto the island. She goes past with Komi in a mini conga train, has her groin pressed up against his rump, walks upright behind him holding onto his hips. They pad around the island together and disappear out of view.

'Hello,' I hear Hannah shout from the gorilla kitchen. 'Do you need a hand?'

'Just finished.' I glance at the time on the monitor. Ten-past eleven. In 21 hours' time I'll be starting the clean all over again.

Diet prep is next. Each gorilla has two buckets of food. Big black 10-litre plastic buckets with wonky handles. Jock's get filled to over-flowing. Sal, on a diet, has a meagre selection in the bottom of each bucket. They each have a name tag for ease later on. I let my subconscious direct, each action taken on autopilot, the diet prep slot honed to the most efficient actions, so I can get back to Gorillas in time for the public talk. The manic pace eases off once the talk is complete and I know I'm nearly there.

Eight cucumbers, 30 tomatoes, 12 yellow peppers. A couple of kilos of carrots and parsnips, courgettes and countless lettuces – iceberg, leafy lettuces, romaine lettuce – leeks, onions and a box of sweet-smelling basil, the food for the public talk separated in its own plastic crate. I add a tray of natural yoghurts, half a bucket of Old World monkey pellets and a coconut each for tea.

Bob helps me lug the trolley of diets back to the gorilla house as the clock chimes 12.15 and we manoeuvre between pushchairs and mobility scooters. The packs of lads from one of the school groups we've got in today are all wired and on the prowl. A rival troop has already been chucked out of

Twilight World for jumping out on each other, screaming and crashing about in the dark.

The lads spot some girls from the rival school and they're suddenly light on their toes, touch each other's arms, crane their necks and stumble against the double line of little kids, each holding the hand of their walking partner, all wearing hi-vis jackets and overlarge sunhats.

We unload into the gorilla kitchen, stack all the diets away with time enough to chop one bucket of vegetables into throwable chunks. The food that Bob will throw over the moat as I deliver the public talk. Once the food is chopped, I relax. The manic pace of the morning routine has paid off and I'm exactly where I need to be.

The gorillas know what time it is and keep an eye out for us through the public windows. As we walk through the house, Bob laden down with a crate of food and me with a microphone and headset around my neck, the gorillas all shoot outside and take up their regular positions around the island facing the podium.

The speakers crackle and the microphone flashes *low battery*. I blow into it a couple of times as we work our way through the crowd. Bob leads the way and mounts the podium with the chopped food. 'Ready to go?' he says.

I nod and stand on the lower step. 'Afternoon, everyone.' The microphone squeaks. 'Welcome to the gorilla talk.' I fiddle with the wires to the headset. 'My name's Al and throwing the food over for the gorillas today is Bob, and we're two of a team of keepers that work with the mammals here at Bristol Zoo.'

The crowd applaud as Jock catches an iceberg lettuce, hurled over by Bob.

'I'm going to start by talking about Jock, our silverback male. You can see him sat here at the front.'

I scan both the crowd and gorillas as I talk. Romina, the perfect fan of hair around her head, sits to one side and pats her belly, on the opposite side of the stream from Jock. Behind her in the wooden shelter is Sal, lying face forward, propped on her elbows, in her pile of straw, Kukena next to her, hopeful eyes on Bob as he sends over some chopped carrot.

'Probably the first thing you notice about Jock,' I hear myself say, 'is his enormous size. He's at least twice, if not three times the size of the adult females we've got here on the island.'

Cheery Komi hovers around behind Jock as does Moki, sitting arms crossed in a piss of course. Kera lurks around the back of the shelter, hopeful but wary.

'Now that size difference…' And I've hit the sweet spot with the microphone headset, voice carries perfectly.

Bob chucks more food.

'The difference in size between a male and female of the same species,' I continue, 'can generally tell you something about the social system that animal lives in.'

Sal waits for food to sail into the shelter, taking advantage of the fanned walls, making it as easy as possible for us bozos to throw it her way.

Romina munches on a bit of leek and Jock picks his way through the iceberg lettuce.

'A gorilla group,' I say, 'is made up of a single breeding male like Jock, a number of adult females, and their combined offspring, which is what we have here. Both Romina and Sal are breeding females. Sal is mum to the adolescent gorilla behind Jock.'

Bob hurls a piece of sweet potato at Cheery Komi to help point him out to the crowd.

'Sal is the mum to Kukena too, you can see her there in the shelter.'

Bob sends a chunk of carrot into the shelter and the tiny Kukena tries to stuff it into her mouth.

Sal sits up and grabs her daughter's head.

'Booo,' a kid cries from the front.

She prizes open Kukena's mouth and devours the carrot herself.

'Sal's on a diet,' I say and the microphone crackles. 'That's why she's stealing her daughter's food… Romina,' I continue, 'is the second breeding female in the group and her child is the grumpy-looking gorilla over there, Moki.'

Moki is making some sort of stand, sulking and ignoring the bits of parsnip and aubergine bouncing around the logs she's sat on. Cheery Komi picks up scraps to retreat with up the bank and Kera kicks the back of the shelter and claps.

The class of young kids cheer and clap in return and whoop when Jock catches another lettuce.

'Kera,' I say, 'was hand-reared at a gorilla nursery in Germany. Kera was born one of twins and sadly her mum rejected them both, so they were sent off to be hand-reared and Kera joined our group here when she was four years old.'

The kids are all pointing and craning their necks.

Moki stalks through the rushes around the front of the island and tugs at the strip of torn pond liner she's been pulling up.

Some ducks flap and quack away from her.

'So the reason male gorillas like Jock are so huge in comparison to the females is that he may have to protect his family group from rival males out in the wild.'

More duck-quacks erupt from the moat.

Jock coughs in Moki's direction.

The kids scream and wail.

Quacks echo across the island.

Moki has grabbed an adult duck by the neck. She stands up on her back legs and swings it around above her head.

The kids boo.

'Save the duck,' one cries from the back.

Jock panics as Moki runs at him. The duck flaps and scatters feathers.

Moki has never caught an adult duck before.

Jock flees off his rock and Cheery Komi leaps up the bank out of his way. Despite the duck terror, Jock pauses to pick his way carefully across the stream, which runs at a depth that would barely cover his wrists. He gingerly waddles across it, giving Kera the chance to dart up the climbing frame, lips pulled back in fear.

Sal sits up in the shelter and Kukena scampers behind her. Romina glances up unmoved.

Jock charges after Kera, but she scurries further up the climbing frame, well aware his fear will be hurled in rage at her. She perches out of reach.

'Save the duck,' the kids begin to chant. 'Save the duck.'

Moki runs in a circle, spins the duck like a lasso and crashes back through the slide into the house with it.

Bob rains tomatoes and peppers around Jock to distract him.

Sal lumbers up to peer around the corner of the shelter, whips up a piece of long bamboo she's been saving and reaches out to fish a stray tomato her way.

Jock has given up on Kera and picks his way carefully back over the stream, worried about Moki, who hasn't yet come back outside.

'I'm going to try and save the duck,' I say into the microphone.

Bob nods and I drop off the podium, sliding between the public, who are all craning their necks to see over each other's shoulders. The duck will be going on a one-way trip up to the vets, no two ways about that.

'I'LL TELL YOU ABOUT ROMINA NEXT,' Bob shouts, as I've abandoned him and taken the microphone with me. I gently push past the kids, who are still chanting, pad through the public area and clink my way through all the different sets of doors to get down into the keeper-corridor.

Moki has let go of the duck and now cowers in a corner, hands over her head and one eye.

I see Jock on the monitor, standing on the upper level above the floor of den two. He barks. A gorilla bark is a call of fear and a warning to the rest of the group.

Moki flinches.

I power up the remote control and begin to shut slides. The two leading outside onto the island clunk closed and the huge portcullis drop slide grinds slowly down from the ceiling.

The duck quacks.

'Come on, you two.' I pick up the white plastic tub of peanuts. 'Jock, Moki.' I rattle the tub as I walk down the corridor and see Jock in the monitor, upstairs above me, already on his way through. Moki sidles past the duck and flees the den.

'You really are vile, Moki,' I say. The drop slide booms shut and I start on the inner slides, to isolate the duck.

Jock clambers down into the basket, coughs at Moki as she comes through.

'Yeah, you tell her, Jock.'

He holds out his long hand, palm up, leans his huge shoulder against the mesh window and stoops his head down to look out at me.

Moki looks hopeful too as I prize the lid off the peanut tub.

'You're not getting any of these,' I say to her and pile a load through the mesh into Jock's palm. 'You're an arsehole, what have you done to the poor thing?'

My radio crackles at my belt. '*You've still got the microphone on,*' Bob informs me.

I grab the microphone out of my pocket, the light gleams green, hold down the off button, claw the headset off and shove it all on the deep windowsill overlooking the island outside.

Shut down the last slide and unlock the door.

The duck's neck is elongated, like a swan, and the top half of its bill has come off. Its webbed feet paddle on the floor. I radio the vets and wrap the poor thing in my fleece.

Bob has finished the talk by the time I get back down and is answering questions; the crowd has begun to disperse.

Kera and Cheery Komi hover around the back of the island and Romina is picking through Jock's leftovers.

A guy with a bright red face stomps toward me as I walk toward the front. He leans forward, almost off-balance in his determination. 'The duck?' he says.

'It's up with the vets.'

'Why the bloody hell do you keep them there?' The vein down his forehead stands out.

'They're wild ducks,' I say, keep my tone straight and even. 'We don't keep them, they like the moat, so they sometimes nest here.'

'Well...' He folds his arms, and his chin is suddenly nestled back in his neck. 'I think it's...' He shakes his head. 'I think it's disgusting, you should be ashamed.'

'Moki does it to get a reaction,' I say. 'That's her motivation, so we don't reward that behaviour, and we're hoping she'll grow out of it soon.'

'Reward it?' the guy shrieks. 'I should bloody well think not.'

'And she was scared of it,' I add. 'When she took it inside, she was terrified, so hopefully that'll put her off.'

'Do you think this is funny?'

'No.' I shake my head. 'These are animals. Sometimes they do things we don't like. I'm sorry you had to see it.'

The school kids all cheer.

I look around and Bob is crouched down with them, handing out strips of zoo animal stickers. He points at a duck on the moat and nods his head my way. The kids all turn to look at me as one.

The guy stalks off and I get applauded by the kids as I walk through them back to the messroom for lunch.

'You told them the duck survived, didn't you?' I say to Bob as he follows me upstairs.

'Course I did. Said you'd just chucked it back in the moat, pointed one out, you know. Them kids were only about five and I didn't swear in front of them like you did. *Moki, you arsehole.*'

'Did you see Kera and Komi doing that conga-line behaviour?'

Bob tears the foil off a Pot Noodle. 'Thought he looked especially cheery.'

Hannah clatters upstairs to the messroom. 'Romina's in den one sweeping up.'

'What?' I look up at the monitor.

'Get her to do the island next,' Bob says and flicks on the kettle.

'There are some cherry tomatoes downstairs,' Hannah says. 'In the fridge if you need to trade for the broom.'

It costs me four cherry tomatoes. Romina, happy with the rate of exchange, pushes the broom back under the door, slightly chewed on the handle but still in one piece. The pegs it hangs from are loose and I make a mental note to sort it out this afternoon.

Hannah and Bob are both looking out of the window when I come back upstairs.

'Look at these two,' Bob says. 'Loving life, look.'

Kera and Cheery Komi are lying next to each other on the roof of the shelter, both with their feet in the air, idly poking wrists. I can't hear them from here, but it looks like they're both laughing.

Months later, when we looked back through Kera's animal records, trying to figure out a due date when she fell pregnant with Afia, today was the first time we entered a note, tagged under *sexual behaviours*. Cheery Komi and Kera's behaviour developed, and they began mating about five months later.

Theoretically Jock shouldn't have let his son mate Kera, but as he'd never known what to make of her, he didn't care. Kera's behaviour still made no sense to him and he had no idea how to treat her.

Kera and Cheery Komi were having a great time and she was getting some social interactions at last. The hope was that if Kera did go on to conceive, even if the baby was fathered by Cheery Komi, her place in the hierarchy would improve, as Jock would see her in a new light.

That was the hope.

Chapter 4

Good News for Ducks

Gorilla husbandry relies on the relationships we build with the animals, which allows us to train them. Training centres on potential veterinary procedures and general animal management, from weighing them and checking them over for injuries, to taking their temperatures with an ear thermometer and injecting them by hand, through the mesh, when they need anaesthetizing.

The first gorilla I had any success in training was Romina. Romina was kind, there's no other way to describe it. I'm putting a human emotion on an animal, but she was genuinely kind to us and the rest of the gorillas. Romina had cataracts for the first 19 years of her life and was given replacement lenses when she came to Bristol, a world first for the western lowland gorilla. She became fascinated with the tiniest of details out on Gorilla Island, all the little things she'd never seen before, and would often be seen with her arse in the air staring intently at an ants' nest in the grass. She was a bit younger than Sal, with way fewer battle scars and a long, smooth face, the crest of fur on the back of her head always a perfect fanned sweep of grey, tinged with red. During the afternoon feeding routine, Romina didn't piss about like the others. She would wait patiently to go into her den for tea and ignore the younger ones as they hurtled about or hovered in slides. She would sit, calm and good natured, and never tried to grab at keepers through the mesh or stab you with sticks. She had doted on her daughter Moki when she was little (before she entered the troublemaker phase), never left her behind, supervised play between her and Cheery Komi when they were both balls of rolling, laughing fur. Romina had gone off Moki, as had everybody else, a couple of years ago and now much preferred playing gently with Kukena, Sal's youngster. Jock preferred Romina to Sal, not that she ever figured this out or exploited it. Romina even tolerated Kera.

That's not to say she didn't have a few quirks of course.

When snow covered the island one year, we offered the gorillas baked potatoes for their tea as a novelty food item, to help warm them up.

Romina went berserk.

The first glimpse, or maybe just sniff of her baked potato, nestled amongst the lettuces in her food bucket and she went into a total frenzy of desire. Sheer, naked want. Coughing and pounding on the keeper-door, doing full backward cartwheel leaps and biting at her own wrists.

Her sense of smell seemed more advanced than the others, maybe something to do with her eyesight having been so poor at the beginning of her life. One of her favourite things was to hang out with you when you cleaned the gorilla drains in the keeper-corridor.

There are jobs you do as a zookeeper that are nothing short of disgusting and the gorilla drains was one of those. Thursday was designated drain-clean day and required four large black bin bags – three to lie on, one to wrap around your arm, up to the shoulder ideally – and a plastic slop bucket. The big square iron drain-cover in the keeper-corridor was removed and propped against the wall to drip black sludge. The drains had a wire trap inside, clogged with compacted wood-wool and gorilla shit, all the collected gunk from a week of hosing the gorilla enclosure. The stench, once disrupted, stayed with you for the rest of the day. You always ended up with grimy arms and once the mesh trap was emptied, you had to reach your arm down a sort of U-bend. Eggy, and if you weren't careful, wretch inducing, almost as bad as otter faeces. A truly vile smell that Romina absolutely loved. She'd wait expectantly on a Thursday afternoon at the keeper-door and post through clumps of wood-wool and pieces of browse, twigs stripped of bark, all to show her willingness for a system of exchange, ever hopeful you'd dish out some slop from the bucket in return. She would grumble encouragement, shove her face to the floor and squint through the door once you got going.

There was a gorilla drain in the keeper-corridor outside the isolation den. We needed daily urine samples from Romina to check her pH levels and were managing to grab the odd one where we could.

I trained Romina to pee on command in the isolation den and first captured the behaviour on a drain-clean day. Gorillas are clever and associate verbal praise and rewards with the behaviour they'd just performed.

I lifted the grate outside the keeper-door and Romina immediately came into the isolation den to investigate. She was keen to be shut in too, a guarantee that if today was finally going to result in being offered a handful of drain sludge, she wouldn't have to fight over it with anyone else.

I continued to empty the drain and waited for her to pee. Eventually she did and I rewarded her – not with sludge – but with some grapes I'd stashed on the shelf. I let her back out again, so I could get into the den with a syringe for the pee.

It took Romina a week to realize when I stood next to the isolation den keeper-door, all she had to do to get a grape was come in, pee on the floor and walk out again. She'd wait patiently for us to clean the den first, so we could collect the urine after she'd walked back out, from as clean a surface as possible, and she knew we'd do it at 10.45 every morning.

Eight months had passed since Moki swung the duck around her head and it was now time for her to leave the collection for a new home with a different troop of gorillas as a breeding female. Happily for ducks everywhere, her new enclosure did not include a moat.

Gorilla moves, where possible, are timed to coincide with the natural dispersal age of wild animals. As male and female gorillas grow up, they begin to hang about at the edge of the group. The males will eventually head off on their own or, if there are a pair of male siblings, leave together. A male gorilla will either live alone or become a member of a bachelor group of blackbacks. It can take a further ten years before they become fully grown silverbacks, when they start looking for a family group to take over.

The young females wait for a different family to wander by and if the silverback seems okay, they sneak off, leave their parental family behind and join the new group. Much like your kids, there is the gradual absence as they get into their teenage years, the push-pull and eventual leaving home. Nobody wants to live with their parents after a while; instinctively, we're all driven to disperse.

Moki had reached the right age to be leaving the group and it was clear that Cheery Komi would also have to move to a new home soon, as he'd mated Kera. His bond with Kera had accelerated his development and opened a door for him into a wider world. Jock was still tolerating Komi and was playing with him each morning, wrestling and laughing, enjoying the tactile strength of his son. Jock loved to play with his kids but was usually forced to be gentle and restrained. Wrestling with Cheery Komi, as he beefed out, was way more fun. However, we knew the clock was ticking on finding Komi a home before life got less cheery.

The European captive population of western lowland gorillas is managed in a collective way across the numerous zoos that house them. A common misconception is that zoos own their animals and sell them to other collections. The reality is that zoos work together and move animals around to maintain a level of genetic diversity within the captive population as a whole, ideally at a comparable level to their wild counterparts.

The reason wild gorillas leave their family group, why any primate species disperse, is an instinctive drive to avoid inbreeding. The breeding program manages the genetic diversity of the captive species and recommends moves to other collections at the right point in the animal's life.

It can take time to find a collection with space enough to take an animal, particularly one that lives in a social group. However, the time it takes allows you to train the gorilla for 'hand-injection', meaning the animal is trained to let you inject it in the shoulder to anaesthetize it, as before it leaves, it needs to undergo a full pre-export health check. Ideally you also train it to walk voluntarily into its export crate.

Bob and Hannah had been encouraging Moki into the crate and shutting her in with a big load of fruit, gradually increasing the amount of time she sat in there before letting her out again. She'd had her pre-export health check and today was the day of her export.

Bob swings open the metal safety door and I can see he's tense, his smile tight as he tries to keep his bubbling misery about Moki's departure at bay.

The heavy familiar stench of gorilla hits me as I follow him down the keeper-corridor.

'How is it up in Twilight World?' he says over his shoulder.

I'd moved section by now and was up in the nocturnal house again but was keeping my hand in at Gorillas on the weekends.

Sal leans against the door in her usual spot and grunts when she sees me.

I grumble back at her. 'It's good,' I reply to Bob. 'Always nice to get back down here for a few hours though.'

'You miss them, don't you?' Bob's voice breaks and he sniffs. 'When you don't work them every day.'

'Hi,' Hannah calls. She stands at den one keeper-window, Romina in the nesting basket on the other side of the mesh.

Romina purses her lips.

'Hey Romina,' I say.

'How's life on small mammals?' Hannah says.

'I just asked him that,' Bob snaps.

'Yeah, it's good, different smell.'

Hannah and Bob have just cleaned the house. They're dark with sweat, polo shirts stuck to their torsos, trousers soaked, reeking of gorilla piss and the drains, those deep, dank drains disgorging disrupted shit stench.

'We need to get a piss sample from Kera.' Bob takes a sample pot out of his pocket.

Hannah nods and points at the chart on the wall, the vinyl day-by-day calendar. 'She hasn't come into oestrous.'

'Cheery Komi might've done the business,' Bob suggests.

'We need to do a pregnancy test,' Hannah adds. 'So if you see Kera wee anywhere, we need to get it if we can.'

I nod. 'Okay. Where do you want me?'

'Jock down the end, please mate,' Bob says. 'Right Hannah?'

'Train Jock for as long as you can,' Hannah says. 'Really drag it out. We'll let you know when Moki is shut in her crate.'

'And off she goes,' Bob sighs. 'Farewell, you miserable, duck-murdering cow-bag.' Tears fill his eyes.

Hannah squeezes his shoulder.

'At least the journey time isn't that bad,' I say.

'Exactly.' Hannah rubs his arm. 'You can visit her.'

'Guess we should get on with it, then?' he mutters.

Jock spots me walking past the keeper-door and knows it's time to train. He strides across the floor toward the far window and the heavy oak shelf that acts as his training station.

I pull on a pair of surgical gloves and clip the training pouch to my belt, stuffed with tiny pieces of mango and apple, a big handful of peanuts, some chopped pear and grapes. This training session needs to go on for as long as it takes for Hannah and Bob to get Moki in her crate. I'm going to run through his full range of trained behaviours.

I gorilla grumble as I approach, so Jock knows I'm there, and he squashes his massive head against the mesh to see me coming. His eyes are fixed on the training pouch. Sweat lines his long upper lip and he opens his mouth wide before I've asked him to do anything, offering me the chance to check all his teeth and gums. His tongue is fat like my hand and so long, stretching back down his pink and black throat. His canines gleam and he's got strings of leek wrapped in his back teeth. His onion breath rolls over me.

I take up my place in front of Jock. 'I'm good to go,' I call down the corridor to Bob.

'Really drag it out,' Bob says. 'We'll tell you when she's in.'

I wait for Jock to close his mouth and then make the hand signal for him to open it again. A capital 'L' with my thumb and forefinger. He yawns wide and waits, pushes his lower lip through the mesh.

'Good,' I say and post a piece of mango into his mouth.

Bob has laid out the training targets and other bits and pieces for me. The ear thermometer and the dreaded secateurs.

I pick up the first hand-target, which is like a wooden mallet, a short length of broomstick for a handle, attached to a block of wood that acts as a stopper. The block at the end stops Jock pulling the target through the mesh window as he holds onto the broomstick handles.

'Hold.'

Jock closes his grey fist around the handle.

'Good.' I pass him the second hand-target. 'Hold.'

He grips it and leans his head to the mesh so I can post through another piece of mango. My fingers are now safe from being grabbed, as Jock's hands are busy holding the target handles.

'Good,' I say and feel his lips brush my fingers through the surgical gloves.

His frame blocks the keeper-window. The hair across his shoulders and down his long arms is a thick coal black; his thighs, back and belly all silvery grey.

I hold up the nasal swab and he tilts back his head and stretches his upper lip. The two giant nostrils flare and he looks down his nose at me, eyes dark, beneath the thick, curled eyebrow ridge.

I place the swab at the base of one nostril and his nose and muzzle quiver, but he holds still. 'Good.' I take back the swab and offer him a tiny square of apple from the pouch.

The slide into the isolation den grinds open. We can't see it from here. Bob will be calling Moki through to the front half of ISO. The export crate, its ratchet strapped to the dividing door, will be in the back half of the den and ready to go.

Jock turns his head at the noise and lets go of a target. The fur on top of his head above his brow is short and soft, tinged red from the sunlight that falls through the windows. The crest of the back of his head starts in line with his dark grey ears; long upright hairs that merge into the grey silver across his back. All his hairs, short or long, quiver and his chest muscles flex as he coughs in the direction of the slide.

'Jock, you're okay. Good lad.'

He shuffles on the training station and lowers his head to look back at me.

'Hold,' I say and offer him the target again.

He takes it and I feel the strength in his grip tense, because he knows something is up, which of course it is. We're loading his daughter into a crate next door and he'll never see her again. She'll vanish as far as he's concerned, and I rehearse again in my head what we're doing. Moki is the right age to leave the group. Despite the fights she causes I'll still miss her and Bob will be devastated, but I know we're doing the right thing. There is no doubt that I'm complicit in preventing Jock from seeing what is going on two dens away, exploiting our relationship to trick him into believing everything is fine.

'Foot,' I say and Jock shuffles again on the platform, so he can push the soles of his feet against the mesh. A gorilla's back legs are shorter than ours, in comparison to body shape. Their hips are different too, as they remain quadrupeds, so this mammoth animal can sit comfortably with the soft soles of his feet jammed up against the mesh, the 'giant teddy-bear pose' as Hannah calls it. He holds both hand-targets and reaches forward with his lips for rewards. I touch each toe in turn and poke the soles of his feet. His skin is still dry on the left side of his foot, but it's better than it has been. All his toenails are okay, if a bit shit-caked.

'Good.' I put a monkey nut into his mouth.

He splits the shell with his teeth and moves his mouth with the usual practised ease, the cracked shell discarded off his lower lip.

'Give,' I say and he lets go of the first hand-target. I place another monkey nut.

His eyes flick from the targets to my hands. Beads of sweat coalesce and run from his wide nostrils. He leans forward, ready to respond.

I push the target handle back through the mesh, in line above the other one.

'Hold.'

He shuffles his bulk of muscle and fur around on the training shelf, scatters monkey-nut shells onto the floor and takes the offered target handle. Both his giant hands are one above the other and he leans his huge shoulder against the mesh.

'Good.' Monkey nut.

I hear Hannah's voice down the corridor say 'Good girl, Moki, good girl.'

The slides grind and I know Moki is now in the isolation den.

Jock turns his slab of a head and coughs at the noise, lets go of the bottom hand-target and cuffs a split monkey shell off his lips.

'I'm heading out on the island,' Bob calls down the corridor.

'Jock,' I say, quietly. 'You're okay.'

His eyes dart back to me and the training pouch of treats. He presses his jowl against the mesh to peer past me down the corridor as he hears the clink of Bob's keys.

I gorilla grumble at him. 'You're okay,' I repeat, tone low and calm.

The island door creaks open. Bob will be out there hurling all sorts of treats and distractions about; seed mix, monkey pellets, chopped browse and herbs.

I offer Jock the hand-target again. 'Hold.'

His massive fingers fold around the handle.

'Good.' I deliver a nut with one hand and pick up the syringe with the other, needle in place but covered with the plastic cap.

I show it to him.

His eyes clock it and I know he recognizes the syringe. His association with the needle on the end is that sometimes, very rarely, it hurts.

Occasionally, we need to inject the gorillas to knock them out for health checks or dental work. Jock is suspicious but still willing to risk it, and he can see I have left the cap on.

'Arm,' I say.

He screws up his courage and presses his gigantic shoulder against the mesh.

I press the capped needle against the tense muscle as he leans into it, just as I would if I were anaesthetizing him, count to five, the time it would take to inject the drug, and take the syringe away.

'Good, Jock, good.' I deposit three nuts in a row into his eager mouth.

I hear the grind and clunk of the isolation den slide and hold my breath.

Hannah's voice echoes out of my radio. *She's in. Crate secured.*

Jock glances over his shoulder.

'I'm sorry, mate,' I say and reach into the training pouch for a piece of chopped apple.

The island door clangs shut and I hear Bob in the corridor. 'Everything is good to go outside,' he says. 'I'm heading out to the podium with coconuts. I'll radio when we're ready to let them out.'

Jock watches as I put the capped needle down and pick up the ear thermometer. He immediately presses his smooth, dark, almost delicate ear against the mesh.

'Hold,' I say.

He shudders but keeps his nerve, waits as I place the thermometer in his ear and count to ten. I'm not taking his temperature today; sometimes the beep freaks him out.

I see Bob over Jock's shoulder through the huge glass windows as he strides through the public area, a large cardboard box of coconuts in his arms.

'Good.' I put the thermometer down, load two grapes into his mouth.

He gulps them down.

'Give.' I take away the first hand-target.

He shuffles to face me, nudges the mesh with his bowl of a hand.

'Good.' I push some chopped apple through into his waiting palm.

He licks the tiny pieces off his hand with his huge tongue. 'Give.'

He lets go of the second target as I touch it, cups both hands for chopped pear, gulped down in one.

'All finished.' I show him my empty hands, make a cross with my forearms and leave the training window. 'Sorry mate,' I whisper again.

Bob's voice comes out of my radio. *'Let them out,'* he says.

Hannah has the remote.

The slides out onto the island all open and the group rush out, excited by all the treats Bob's scattered. They stalk across the island to manoeuvre themselves into the best coconut positions. Jock clatters outside, eyes fixed on Bob up on the podium.

'All out,' Bob says through the radios.

The slides grind shut.

None of the group have noticed yet that they're a gorilla short.

I take Bob to the pub after work.

Hannah has gone with Moki to be a familiar face when she arrives at the new zoo and is staying for a week to help with the introductions. She's given me strict instructions to be with Bob until he breaks and then make sure he's okay.

I know Bob wanted to be the one to go with Moki. He demolishes two warm cans of beer on the way.

The beer garden is almost empty and I lead him to a table down the back by the ivy-covered wall, beneath the canvas cover. It ripples in the wind and patters with rain. The tables and benches sit in trellis alcoves, strung with fairy lights. Beer keg stools are scattered about. There are piles of blankets for when it gets too cold and orange glowing space heaters.

'Wish I'd gone with her,' Bob says for maybe the fourth time.

'I know, mate.' I place our pints on the table. 'Hannah knows what she's doing, though.'

The table creaks as he slumps onto the bench. 'You just feel such a prick, though.'

'Come on, mate.' I hear a similarity in tone to my Jock training voice.

Bob throws his tobacco and filters onto the table and reaches for one of the circular glass ashtrays. 'We tricked her at the end of the day.'

'You know the alternative would've been horrendous.'

He nods and slugs his pint. 'Yeah. Old Jock could have keeled over with a heart attack.'

'Exactly.' I know Bob needs to talk it out, reassure himself that we've done the best for Moki and the group as a whole. 'And she'd have gone by now in the wild,' I say. 'Dispersed, joined another group.'

'Not with that with grumpy face.' Bob digs a stick of cigarette filters out of the yellow box with dirty fingernails. 'Who'd choose Moki?' He pushes a filter out of the cellophane tube. 'Frowning… moody… Did you see Jock out on the island looking for her?'

I shake my head.

'Poor sod.' Bob slithers out a Rizla paper and places the filter at one end. 'He was out there calling for her.' His voice catches in his throat. 'Looking about all over the island. In and out through all the slides.'

'Come on, mate,' I say again. 'It went really well. If the group knew what we were doing, they'd have gone ballistic. Kera would've got a pasting.' I turn my glass on the soggy beer mat. 'They'd be terrified of us too, lose trust in us.'

Bob sniffs. 'I'm going to miss her, though, that's all.'

I raise my glass. 'Here's to Moki.'

'Kera must be dancing the fandango right about now.' Bob clinks his glass against mine.

'And the ducks,' I say.

'I'm going for a piss,' Bob states and the table rocks again as he pulls himself up. 'Starving hungry. Are you?'

'We can order a pizza,' I offer.

'At them prices?' He shakes his head and plods up the beer garden, eyes a discarded plate with pizza crust on it.

Bob's phone buzzes and rattles on the table. I can see 'Hannah the spanner' flashing on the screen.

I reach out and answer. 'Hello, it's Al.'

'Al?'

'Bob's just gone for a piss. We're at the pub.'

'Any tears yet?' Hannah asks.

'We're getting there,' I say.

'This might cheer him up,' Hannah replies and I hear some sort of Tannoy echo behind her excited voice. 'I got a urine sample out of Kera at the end of the day and took it up to the vets. Guess what. She's pregnant.'

'Amazing,' I say.

'I checked the daily report,' Hannah babbles on. 'We're waiting for the ferry, so I checked emails and it's there on the vet notes. Kera tested positive for pregnancy on one line and then Moki left the collection on the next.'

'How is Moki?' I ask. 'Have you checked her?'

'Quiet,' she says and the Tannoy sounds again in the background. 'I've got to go. Good luck with Bob.' She ends the call and the phone screen fades.

Bob clumps back across the beer garden. He's chewing something and I see pizza dough crumbs in his stubble. He sits back down. 'Leftovers everywhere inside,' he says.

'Kera's pregnant,' I blurt. 'Hannah checked the report when she was waiting for the ferry.'

Bob's face doesn't light up as much as I'd hoped. 'Hope she looks after it,' he says. 'Hand-reared weren't she, no experience.'

The rain stops as we head back toward his tiny zoo flat.

He begins to blub.

I put my arm around him.

His own heavy arm curls around my shoulders and he shudders with deep cries. 'I'll miss her,' he says through snotty despair. 'She's in that crate. Won't know where she is.' He sniffs and whispers 'Moki' to himself.

Chapter 5

The Social Climber

With Moki long gone, seven months later we got a new arrival: Touni. The move was recommended by the species coordinator, the person that ran the gorilla captive breeding program. One of the key roles that zoos play is to maintain the highest levels of genetic diversity in the species they collectively manage. Genetically, Jock, our silverback male, was very under-represented in the breeding program as a whole, which meant that at the time, he was the eighth or ninth most genetically important male in Europe. It was essential that his genes fanned out across the captive population as his offspring went on to join other groups and they themselves bred, to contribute to the genetic diversity across the species.

I'd been promoted by this point and was at the same level as both Bob and Hannah. I was the senior keeper of small mammals and worked in the nocturnal house with the aye-aye and naked mole-rats, the pygmy slow loris and the huge Livingstone's fruit bats. I continued to work in Gorillas now and again and helped out on any big days that were happening, but generally I was elsewhere.

I join Bob in the gorilla house public area after the morning meeting. Touni arrived from a zoo in France last night but refused to come out of her crate. She spent the night locked in the isolation den. Den one is shut down to provide a buffer zone between her and the rest of the group.

'*Opening the slide.*' Hannah's voice comes out of our radios.

'Copy that,' Bob says.

The hydraulics clunk and groan and we watch the two slides into the isolation den grind open.

'She's there, look.' Bob points.

Touni peers out into den one. Her face is small and neat.

'Christ,' he says. 'They haven't sent us a chimp by mistake, have they?'

'How old is she?' I ask.

'Eight. She's tiny, poor cow.'

Feet sound on the glass above our heads as Cheery Komi runs the length of the huge drop slide, the giant steel frame portcullis which lowers from the ceiling to divide den three from den one and is separating Touni from Komi upstairs.

Touni flinches as he bashes the mesh. She ducks and scampers back out of sight.

Jock is at the dividing mesh on the floor in den two. He puffs up his giant shoulders, his back bowed, a perfect taut curve of silver fur, all his muscles bulged.

'He seems to like her then,' Bob says. 'New French lady.'

'Did you see her handover notes?' I ask.

'Briefly had a squizz yeah.'

'The stuff about her being a climber? Hannah said they made a big deal of it.'

Bob looks at me. 'A climber?'

'That's what they wrote.'

'You'll be telling me she likes peanuts next,' Bob says.

Cheery Komi rumbles across the glass overhead again. Puffed up just like his dad.

'Are you seeing this?' Hannah's voice asks out of Bob's radio.

'Komi's displaying for her too,' I say.

Bob looks up at him. 'Bollocks.'

The slides into the isolation den grind shut again and Touni, safe behind the mesh, takes a peek at Jock across the empty den one.

'I'm popping out,' Hannah's voice states.

Cheery Komi is young. In the wild he'd have no chance of getting anywhere near a female to mate. The silverback males would chase him off and if they got into a serious fight an animal Komi's age would be torn apart. Jock was about twice his size. However, Komi and Kera's behaviour over the months had built toward mating. Jock still treated Kera with suspicion and couldn't care less if Cheery Komi was hanging out with her.

'Puts a spanner in the works,' Bob says. 'Look at Komi strut.'

Hannah joins us in the public area. 'We're going to have to separate Komi,' she agrees.

'Because he's been getting his end away with Kera,' clarifies Bob. 'Thinks he's the breeding male now, look. Bonjour, you foxy Frenchie.'

'That's the nature, nurture debate on show,' I add. 'The way he's displaying.'

'Here he goes,' Bob says to Hannah and I'm not sure if he means me, fearing I'm about to bang on about primate social systems or if he's referring to Cheery Komi, who thunders across the glass again and spins to a halt above our heads.

'Touni's the first new female that Komi's known,' I say.

Hannah nods. 'He's never seen Jock display like this either.'

Bob laughs. 'Can't be bothered to turn it on for Sal and Romina no more. Can't say I blame him, toothless old hags.'

'He's putting on a show now, though.' Hannah refuses to bite.

Jock stalks to the slide into den one for a better look at Touni.

'Trim Touni,' Bob declares. 'That's what we should call her.'

He means that's what he's going to call Touni from here on out.

'Definitely can't risk mixing the three of them together,' Bob adds.

We all nod.

'It's going to make introductions way more complicated.' Hannah looks at her watch. 'We'll only have half the options, if Komi is going to be kept separately.'

Trim Touni came from a very stable family group in France and had witnessed several births, helped out raising kids and understood gorilla social behaviour. She was a social climber and clambered through the group hierarchy with an adept ease, beginning by buddying up with Kukena – Sal's young daughter who was now five years old. Kukena had been terrified by Moki at an early age and by extension Kera and her own brother Cheery Komi, as they were all a similar size. Romina would play gently with Kukena when she was little, but she'd never had a playmate to wrestle with and chase about. Trim Touni, three years older than Kukena, manoeuvred herself into this role to help her entry into the group. She played with Kukena, very gently at first, showing Sal she wasn't a threat, and within a few months had managed to create a power trio, with Dangerous Sal as the leader. She was even confident around Jock, despite the horrible battering he dished out at the end of her first week.

Kera was left back at the bottom of the hierarchy again and now Cheery Komi had been separated, she was alone and began to pluck the fur from her arms and shoulders. We made an effort to spend time with her each day.

Kera often comes back inside earlier than everyone else during the cleaning routine, gathers up armfuls of vegetables and comes to sit in peace with us, rather than the rest of the group outside. It means cleaning the house in two halves, so no matter where we are in our morning routine, there is always a den available for Kera to retreat to. Despite her loneliness, Kera remains cheerful, ever hopeful that she might find her place in the hierarchy and have friends. I feel particularly sorry for her as she'd formed a relationship with Cheery Komi and now of course, that is denied her.

Whenever I work the gorillas I set time aside to give her some social interaction, even if it is with the wrong species.

She pokes her head around the corner to one of the slides that divides us and grumbles. I follow suit, peering through the mesh at her, and she darts to the next slide along and does the same, backward and forward, laughing her deep gorilla chuckle the whole time, and keeps going until I have to stop and carry on with the clean. Kera's clever too. More intelligent than the rest of the group, except perhaps Dangerous Sal. We provide the gorillas with enrichment on a daily basis. 'Enrichment' in zoo-speak refers to anything you can supply your species that enriches their lives, by stimulating behaviours they would naturally exhibit in the wild. The gorillas have an enrichment box that hangs on the keeper-side of the mesh designed to encourage problem solving. The box is a cross between a vending machine and a miniature bookcase; it has four shelves inside, each with a hole at opposite ends. It is covered in Perspex, drilled with rows of holes that correspond to each of the four shelves. We deposit a handful of nuts or seeds through a hatch onto the top shelf and the gorillas figure out how to move them from one shelf to the next and eventually make them fall out of a hole at the bottom of the box into their waiting palms.

Each gorilla has a different approach. Jock reaches his huge fingers through the mesh to rattle the box where it hangs in the keeper area. He shakes it hopefully and occasionally pokes the Perspex with a jagged fingernail, or the back of his massive hands. He eventually resigns himself to the reality – that this thing, whatever it might be, is clearly not for him. Cheery Komi tries to squeeze his tongue through the holes, which is occasionally successful, and then kicks the mesh and shakes the box like his dad. Romina ignores the enrichment box altogether and Dangerous Sal puts her daughter Kukena to work. She sits her down in front of the enrichment box and encourages her to fish out the goodies, as Kukena's fingers are smaller than everyone else's and can just fit through the Perspex holes. Any nuts or seeds retrieved by her daughter, Sal instantly snatches away.

Kera goes straight outside, returns with a short piece of stick and strips one end of bark with her teeth. She pushes each nut one by one along the shelf, dipping the stick in and out of the row of holes in the Perspex and repeats the process along each shelf until the nut falls free. She quickly begins to switch ends with her stick, using the flatter end to sweep a selection of nuts along the shelves in one go and using the pointy end for anything more fiddly. I wonder if her place at the bottom of the hierarchy means she has to be cleverer than everyone else to survive. Her day is fuller than everyone else's in the group, trying to work out ways not to offend anyone.

Regular urine samples prove she is still pregnant with Cheery Komi's baby and her belly is beginning to swell. Our hope remains that her welfare will improve massively with the birth of her child, but her behaviour becomes less playful as her pregnancy progresses and she becomes low and depressed.

Toward the end of her pregnancy Kera gets pre-eclampsia. It kicks off with high blood pressure and if left untreated can be fatal for the infant and mother alike. In humans it is usually picked up early in regular checkups. As we are testing Kera's urine regularly we notice a huge spike in proteins, one of the symptoms. Kera becomes miserable and lethargic and four weeks before Afia's due date, she has an emergency C-section.

An amazing thing about working in a zoo is the incredible amount of goodwill showered on you by other institutions. Specialist vets drop what they are doing to come and help out and two doctors arrive from the Bristol Royal infirmary to assist with the surgery. Afia is unresponsive when she is first delivered, having been anaesthetized along with Kera – the drugs passed into her as they coursed through her mum's system. One of the vet team refuses to give up on Afia, ventilating her, or breathing for her, until the anaesthetic drugs wear off, and gives her life. A tiny, squished face, pale and bony, wet-haired with a plastic-looking rope of white umbilical cord. Her head is smaller than a tennis ball. She weighs 1.2 kilos. A bit more than a bag of sugar. The vet team save both their lives and we are hand-rearing a premature baby gorilla.

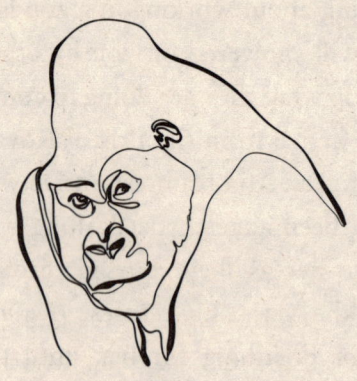

Part Two
Parenthood

Chapter 6

Sick, Horribly Sick

It's a cold Monday morning, with grey clouds overhead and a wind that ushers crisp packets up against the door into the gorilla house. I let myself in and freeze. The veterinary kit is stacked against the wall, orange plastic crates of extension leads and the electric blanket. The anaesthetic machine, on its wheeled stand, with the chamber of tiny plastic-looking balls, and coiled pipes. The oxygen canister propped in the gap next to the gorilla kitchen fridge. I've been off for three days in a row and have missed something drastic. It has to be Kera. She was lethargic last week and nowhere near her usual self.

I grab the treatment sheet clipboard off the hook above the sink, a big new wodge of sheets clamped in the bulldog clip, each one a different medication for Kera.

The chopping board sports two curled banana skins and the plastic screw-topped tablet crusher. The metal jugs we use for their fruit tea are both full of juice and big 50-ml syringes. There are two empty cartons of orange juice in the deep sink and the knife, wrapped with banana string.

Chatting sounds from upstairs in the messroom and I clump up the steps to see Bob with his shirt off, his pale skin blotchy with spots up his back. He turns and his torso has a frown-type look to it, nipples sporting tufty hair like crazy eyebrows, his puckered belly button a surprised mouth.

Hannah comes around the corner in her bra.

'Morning,' I say as I climb up the stairs. 'Why have you both got no—?'

'Skin to skin contact,' Hannah explains, and I glimpse the little gorilla clutched against her chest, furless but for a cap of soft hair above a flat, splayed nose, nostrils thin and drawn across her face, fast asleep.

'How's Kera?'

'Not good, mate,' Bob says.

'She gave birth?' I look at Hannah and the baby gorilla.

Hannah shakes her head. 'C-section.'

Bob lays a heavy hand on my shoulder. 'She's absolutely fucked.'

The messroom is crammed with the entire team, vets too. Everyone is serious and attentive. There is a hierarchy here, much like in any other primate troop. We're led by Michelle, our head vet.

The day is broken down into priorities, the main one of which is to get meds and fluids into Kera. The team disperse after the morning meeting and the vets wish us luck. If we can't get antibiotics into Kera by break time, she'll have to be darted.

Hannah disappears downstairs with the baby to sit with Kera.

'How's she been taking her medication?' I ask.

'Just refused her first lot of anti-b's in some 'nanna,' Bob says as he pulls on his polo shirt. 'Took it off me and dropped them both on the floor.'

Reality jolts in. We're hand-rearing a baby gorilla. 'How's the little one doing?'

'Bit pinchy,' he says. 'You making tea or what?'

I tramp around to the kettle. The tables are strewn with more vet stuff, the dart gun box and a stuffed shopping bag; one of the plastic canvas-style ones, with orange webbing handles, crammed full of all sorts of paraphernalia.

Hannah comes back upstairs. 'Kera's asleep again,' she says and sits at the computer desk, holding the frail baby Afia on her knee.

I try not to stare, particularly if Hannah is feeling uncomfortable sitting there in her bra, but she's way past that. Afia has her hand around Hannah's finger. Tiny perfect fingernails, a sweep of thin hairs across her arms, but grey and bald-looking. A wide mouth and heavy-lidded eyes, the fur on her head parted into two wings. The smallest ears, her skin all loose and baggy.

Hannah plugs in the nebulizer; a dull grey box with an attached gas mask and tube, a machine that turns liquid medication into mist, to be absorbed into the lungs.

'How many times a day is she getting nebulized?' I ask.

Hannah props Afia forward on her knee and the tiny gorilla wobbles and blinks her eyes.

'Twice a day, poor little mite,' Bob says to me. 'Got loads of fluid on her lungs in the birth process.'

Hannah flicks on the nebulizer and it groans and hums. She engulfs the tiny little being with what looks like a giant jet-pilot mask.

'Coffee?' I ask her.

'Please,' she says.

'I'll get on with cleaning this morning.' Bob rubs a gnarled hand across his head. 'Let you concentrate on Kera. She's got sick of the sight of me over the weekend and she ain't seen you for a day or two. Get them meds into her.' He yawns again.

'Sure you don't want coffee?'

'We drunk it all,' he says. 'Only got instant left – rather drink a cup of drain sludge.'

'How long have we got?' Hannah asks.

I look at the dart gun case. The stress of darting a gorilla is huge, not just for the recipient of the dart, fired at close range, but also the rest of the group, who go the very definition of apeshit.

'She's on two different anti-bs and we got to get them in sharpish, so we can get another dose in later.'

As much as I want to hear all about Afia, Kera is the priority. I head downstairs to begin prepping the tablets, a total of 18 all in all, to be ground up and kept in separate empty yoghurt pots, so I know what she's had and what she hasn't.

I begin with peanut butter and mix a dollop with the first set of powdered antibiotics, skim some of the oil from the side of the tub and add a squirt of honey. Stir it and roll four marble-sized balls. No way she'll refuse these, she goes mad for peanut butter and honey is universally adored by primates.

I head into the keeper-corridor and get the icy wash of fear at the sight of the open isolation den keeper-door. Kera is curled up behind the mesh slide in the back half of the den.

I step in.

'Fuck,' I whisper.

She looks dead.

Huddled in a foetal position on the floor, knees up to her chest. Her skin is papery and loose, her fur clumped and dusty-looking.

The bananas Bob tried earlier still lie next to her, ignored. Her eyes are shut, mouth open.

'Kera,' I say and grumble. 'What's this? What have I got for you?'

She lies still and I know she's sicker than I've ever seen her.

I sit cross-legged in front of the slide and surround myself with treats, lay out a stall for her to look at. 'Kera.'

She opens an eye, sunken in her socket and grunts.

'Hey gorgeous.' I give my tone some enthusiasm. 'Look, what's this?' I hold up the first peanut butter ball.

She lifts her head from the floor. Her lower eyelids are both drooped, the skin on her lips dried and cracked, her gums a pale grey.

'That's it, Kera.' I keep my tone excited.

The other gorillas grumble in den one next door. Sal peers through the mesh, delirious with the thoughts of all the goodies being handed out and discarded.

I hear the hose over in den three, Bob already hard at work cleaning.

Kera winces as she leans forward on her elbows and grumbles. She doesn't want me to see the surgical site, has always hidden her injuries, as much from us as the other gorillas.

I offer her the first of four peanut butter and honey balls through the mesh. She leans her face forward to take it and lets it roll off her lip onto the floor. Not even the slightest bit interested in ingesting it, but willing at least to take it from me.

'You've got to take them, Kera.' I think of the dart gun waiting upstairs. I point at the peanut butter ball. 'You dropped it, down there, look.'

She rolls it away with a clumsy swipe of her hand and lies back on her side. I catch the briefest glimpse of her shaved belly, the baggy grey skin and line of stitches. They look intact. The only upside to her feeling this rough is she hasn't started trying to pull them out.

'What about some juice?' I take a big syringe out of the jug. 'It's the good stuff, pure fruit juice.'

She grunts again but squeezes her eyes shut.

'You'll feel so much better,' I say.

The peanut butter ball is in reach with a piece of stick. I fish it back through the mesh and head back out into the kitchen.

Bread. Sandwich time. The restaurant has given us some bread and I squidge the honey and peanut butter balls onto a slice, squirt on some more honey and make a sandwich, cut it into four squares and take the crusts off.

Kera slowly pulls herself up to sit on her haunches and grumbles. I get another brief glimpse of her surgical site but don't stare in case she backs off again and post the first quarter of the sandwich into her mouth.

She immediately spits it into her hand.

'Eat it, gorgeous,' I encourage.

She peels off the top layer of bread and peers through her droopy eyes at the peanut butter, antibiotic-laced sweet delights. She pulls a smear with her fingernail and sniffs it suspiciously, then wipes it off on the floor.

Her whole belly is shaved bald from the operation. The stitches are still all in place and these are an additional line to protect the actual sewn surgical site underneath. She shivers, hunches and yawns again, her pale lips lined with splits.

Looks like the antibiotic's smell is offending her.

'How's about the pain relief, then?' I ask her and head back out into the kitchen for the tramadol yoghurt tub. The fridge is stacked with every sort of goody I can ask for. Some legend from the team has stocked it for us. Yoghurts, grapes, bananas, kiwi fruit. More bread, a couple of rolls from the restaurant and some cold cooked potato.

'Shouldn't have used honey straight away,' I mutter. Rookie error. I've burned my bridges with both peanut butter and honey in one go.

Tub of strawberry jam. 'Yes.'

I make up two rolls and spread the ground-up tramadol in some margarine, liberated from the lunch fridge upstairs in the messroom. I take the tub of strawberry jam with me and sit back down in front of her. She's almost asleep again; still sat upright, but her head nods.

'Kera,' I say quietly and her eyes open. 'What's this?' I pop the plastic lid off the industrial-sized tub of strawberry jam.

She grumbles.

I hold up the first half a roll and dunk a spoon into the jam.

She grunts.

I squidge the jam on top. 'You want it, Kera?'

She takes it with her lips and I try not to hold my breath, pick up the second piece of roll, act normal, relaxed.

She gulps it down, her eyes fixed on the spoon. I post the next three pieces of roll into her in quick succession, washed with a glow of relief. First set of meds on board and I haven't a clue what the time is, don't dare to stop feeding her to check, but I must still have time before the vet team turn up to shoot her with the dart gun.

Bob appears at the den door. 'How's it going?'

'She won't touch the anti-bs,' I say over my shoulder, 'but she's just had some pain meds.'

'Good one, Kera,' Bob says. 'I've whipped her up a Complan.' He brings in a chewed measuring jug, full of garish pink protein slop.

'Can you ask the vets for some more anti-bs and grind up another dose for me?'

'Yes, me lord. Anything else?' He's trying for levity, but I know how worried he is, see it in his eyes and the way his shoulders are hunched.

'Coffee maybe?'

'That instant crap?'

'I'm going to stay down here and keep trying.'

'Try cutting the sandwiches into triangles next time.' He squelches off in wet wellies and I hear him stomp up the stairs to the messroom.

I've got an empty yoghurt pot, like a shot glass, and try her on some Complan. She slurps it down through the mesh.

I take the measuring jug out to the kitchen, so she can't see me pour ground-up paracetamol into the warm pink sludge, stir it up and head back in again.

She takes it. Takes the lot, pushes her lower lip through the mesh, so I can gently dribble it in. I feel like I'm getting somewhere, but it's the antibiotics that are essential.

I follow the Complan with some fruit tea and she pokes the yoghurt pot away with her chipped fingernail.

'Juice, then?' I offer her one of the 50-ml syringes of orange juice and she takes it reluctantly. Normally she'd clamp the end of the syringe in her teeth and pull the whole thing through the mesh, to destroy at her leisure, but not today. The food left in with her last night lies untouched. A gorilla's digestive system is complicated and relies on leafy vegetables to keep everything working. Treats are good for getting meds in, but a gorilla can't live on fruit and bread and she needs an abnormal amount of fluids.

One step at a time.

I bring in a selection of vegetables and add them to the overall display, but Kera has had enough, turns her back and curls into a ball to lie on her side, her head on a clump of wood-wool as a pillow.

I look at the time. Pain meds should kick in within an hour or so, which might mean she picks up. That'll leave me all of 10 minutes to get the remaining meds into her.

'How are you getting on?'

I look around and Hannah stands at the door, Afia cradled in her arms. 'Who's this, Kera?'

Kera grunts but doesn't lift her head.

'Thought I'd do a milk feed in here, see if that perks her up at all.'

The miniature gorilla is wide eyed.

'What's the plan?' I ask.

'The Higher-ups are having a meeting with the species coordinator later.' Hannah has a baby bottle of milk with her and sits next to me.

'Would they send her off to the nursery place, where is it again?'

'Germany,' Hannah says. 'I'm not sure. There's a discussion about doing it differently. If we can get her back with the group, she'll grow up as a functioning gorilla.'

'Back with Kera?'

I know Hannah will be thinking the same as me. Kera is at death's door, but neither of us want to say it out loud.

'We were chatting about timelines with the vets,' Hannah says and stares down at the baby gorilla in her arms. 'It might be we use a new method of hand-rearing. Keep her with us twenty-four hours a day until she's able to come to the mesh on her own for a feed, but once she's at that point we reintroduce her.' Hannah shakes some drops of milk onto her wrist. 'If we can get them back together before she's a year old, she won't remember the hand-rearing process, she'll be able to witness gorilla social behaviours as she grows up. We'll spare her all the issues Kera's had.'

Afia's fingers are delicate, she's all limbs and Hannah's hand fits around her shoulders.

'How long will that take?'

'Six months at least,' Hannah says. 'And we think she's four weeks premature too, so more like seven or eight.'

I watch the feed, slow and gentle. Hannah winds the tiny animal, gently tapping her back with her fingers. 'Who's this, Kera?' Hannah says.

Kera stays motionless and we hear her stomach bubble and gurgle.

'I hope she doesn't throw it all up,' I say.

'It's a nightmare, she had milk come out of her nose last night.'

I don't tell Hannah I meant Kera. I know her attention has focused on the skinny creature in her lap.

'What do you do if she inhales it?'

'Hold her upside down,' Hannah says.

Kera remains flat.

Hannah heads upstairs once the feed is complete and I hear the vets arriving, with the jangle of keys and hushed talk.

Bob appears. 'How you doing, mate? Forgot to bring your coffee down.' He has my red mug in his hand. 'She taken them?'

I shake my head. 'She's been flat out for an hour.'

Kera lifts her head at the sound of his voice.

'I'll let the vets know,' he says.

Kera rolls over onto her front, head facing me, and peers up at my coffee cup. I sip it and she watches.

I sip again and she grumbles.

I unclip my radio. 'Fourteen to twenty-seven vets.'

'*Receiving,*' one of them answers.

'I've got a med question. Do you mind popping downstairs? Don't come in, though.'

I hear feet upstairs and Michelle, our head vet, creeps down the corridor and hovers out of sight at the door.

The vets have a rough time with the gorillas, particularly as Kera had to be darted and taken to clinic on Friday. They all associate the vets with bad things and can't stand the sight of them.

'Can Kera have her anti-bs in my coffee?'

'She can have them in a gin and tonic if it means we don't have to dart her.'

Chapter 7

Running Out
of Options

Kera remained in the back half of the isolation den, lips pale and cracked, her skin baggy and dry, eyes sunken in their sockets, her coat scraggy and dull. Work went into one of the all-consuming phases. I lost track of what day it was and dreamt about gorillas each night, haunted with crushing disappointment as Kera struggled to recover.

Her daughter Afia had been four weeks premature and both Hannah and Bob were hand-rearing her. She was doing well and could push herself up on her stringy arms. The fur across her wiry body was beginning to come through, the shock of thick black hair on top of her head now stuck up straight, giving her an expression of extreme shock.

Trim Touni, the social climber, was fully integrated into the group, playing number two in Dangerous Sal's power trio. Romina confused Touni, her kind and gentle nature baffled her, and they mainly ignored each other. Jock was happy to have Kera out of the way, but as Cheery Komi was living separately too, in case they fought over Touni, the cleaning routine became horribly complicated. We had covered the mesh on the two giant drop slides, the portcullis barriers that descend from the ceiling to shut off den three, and Komi lived in there alone, unable to see his family, but still aware they were all there. However we did now have a date for his imminent export.

As Kera remained flat the vets stepped in to knock her out again and discovered she had severe anaemia. Her body was breaking down her own red blood cells. If she were a human, she would have received a blood transfusion.

On one of those morning repetitive cleaning routines I notice a tiny wound on Kera's leg, small and dry, the size of a fivepence coin. I watch in horror as she pokes at it with her nails and then peels off a long strip of papery skin. Dangerously anaemic, she has developed bedsores, and any additional injuries or any more blood loss could be fatal. The wound itself doesn't really bleed, leaving an orange sore beneath.

We create a soft foam bed for her to lie on, with fitted sheets. This is an animal that eats binbags given half the chance and tear sheets into strips. She falls onto the bed with a groan and lies still.

It is clear the time has come for the final level of veterinary care. Options are discussed and all hope falls on Cheery Komi, due his pre-export health check, before he moves zoo. As we would be anaesthetizing Komi, it is suggested we take blood from him to give to Kera, alongside the iron supplement given to Jehovah's Witnesses in need of a blood transfusion and a drug to stimulate red blood cell production. A procedure to save Kera's life.

We gather in the messroom to sign the risk assessment. The tables are covered in vet equipment, folded open plastic toolboxes, full of syringes and sample pots. A yellow sharps bin, the intubation kit, dart gun and stretcher, boxes of disposable gloves, surgical masks and the anaesthetic machine balanced on its precarious wheels.

Gorilla general anaesthetics, or in zookeeper speak 'knockdowns', happen infrequently, one every couple of years maybe, but by this point we felt like we were doing one a week and the blood transfusion meant two in quick succession.

There is an atmosphere of unity between us all, a collective support and welcome air of confidence and optimism that emanates from the vet team. Their professional dedication and skills are insane. I get an inspirational rush working alongside them. You share crazy heights of emotion and intense hands-on animal action; holding the jaw of a lion open, hands around the lower fangs, as a vet reaches her arm down its throat. Restraining sloths as they claw their way out of rubber transport bins and long late nights, where you sit together quietly hoping your animal will come round from anaesthetic. You fight alongside them to administer emergency care and they gently support you if things go badly and your animal dies.

'Morning everyone,' Michelle says. 'I'll be leading the procedure today.'

The team falls quiet.

Michelle is our head vet and for a dangerous veterinary procedure like this, she is our silverback. We, her group, hang on her every word. Along with me and Hannah, we are joined by Teresa, the vet nurse, Scott, who has come down from Twilight World to support, and two further vets in green scrubs.

I have complete trust in Michelle. She instils reassurance and leadership and we will all put ourselves in danger if she needs us to.

'We're going to be knocking down Komi for his pre-export health check,' Michelle says. 'But as Kera is in the den we'd normally use, we're going to do it upstairs in the end feeding den.'

I nod along with Hannah and Twilight Scott, who's here to stand guard on one of the keeper-doors. If Komi wakes up and tries to escape, it's Scott's job to pull the door shut after we've all fled past him.

Bob is off downstairs with Afia and Kera.

'That means,' Michelle continues, 'the anaesthetic machine is going to be outside in the corridor, so Al—' she smiles at me '—has kindly volunteered to sit behind Komi on the table and keep the mask in place.'

I love a hands-on role in veterinary procedures.

'Hannah will aim to inject him by hand and once he's down we'll need to intubate him, before getting him up on the table.'

Hannah's smile is tight.

'We're working in a smaller space than usual,' Michelle continues, 'so if you don't need to be in the den with us, then wait outside. We'll follow the pre-export list and check his heart, but we're also taking blood from him today, to give to Kera tomorrow.'

We all nod.

'The risk assessment is going around,' Michelle says. 'This is a dangerous animal. As ever, if I say get out, everyone leave.'

I give Teresa a hand to cart more bits and pieces up the stairs. Teresa has always got a story about a jaguar in a tea chest bouncing across a zoo, or her bathtub at home being used to house recovering terrapins. She has worked as a vet nurse her whole life and is another person I trust implicitly and would do anything for.

'Are you feeling happy about tonight?' she asks me.

Tonight I am taking Afia home for the first time, officially joining the hand-rearing team. 'Yeah,' I say. 'Worried about aspirating her during a feed.'

'Just take it nice and slow,' Teresa says. 'She'll learn herself eventually, how to suckle from the bottle.'

'Worried about Kera too,' I add, because Teresa is someone you can be open with.

'We'll throw everything at her tomorrow,' she says. 'Get some blood from the Boy Wonder upstairs and send it off to see if it's a match for Kera.'

'When will we know?'

'By tomorrow,' Teresa reassures me. 'If we're lucky, that'll be enough for her to bounce back.'

I know Teresa is honest too. If she thinks there's a chance, that's good news for Kera.

The end feeding den, where we are knocking Komi out, has a table-sized shelf against the mesh. Better yet, one wall runs at a 45-degree angle out above the floor of den three below. Once Komi is unconscious, we can sit him up on the shelf, with his back against the mesh. Gorillas can run into problems breathing under anaesthetic, so propping them up is essential. The downside to the end feeding den is that it is small and the vets, with their raft of equipment, can only get access to one side of him on the shelf.

Hannah injects him by hand, so we're off to a good start.

We all pull on blue disposable gloves and surgical masks and I go in to check on him with Hannah and Teresa after 5 minutes. The upstairs feeding dens all lead off the top keeper-corridor, much the same as the one downstairs, but with more keeper-doors.

Teresa and I wait for Hannah to creep up and check Komi.

The line of feeding dens all interlink and can be used to form a corridor from den three to outside, which meant we could give Komi access to the island without letting him in with the rest of the group. We would shut them all in den two, but it was tragic to watch. Komi would scoot outside and try to catch a glimpse of his family through the windows and hoot.

'He's down,' Hannah says.

Teresa and I join Hannah and I pluck a broom from the tool rack.

Komi is slumped against the wall.

'Well done,' Teresa says. 'That was quick, let's give him a check.'

I insert the broom handle through the mesh and give Komi a poke.

He doesn't react.

I poke him again and Teresa nods.

'I've got a thin piece of bamboo,' Hannah says.

Teresa takes it and threads it through the mesh too, uses it to gently prod Komi's closed eyelid.

He still doesn't react.

Hannah lifts her radio. 'Fourteen to twenty-seven Michelle. He's down, over.'

Michelle comes in with Twilight Scott, who has the canvas stretcher under his arm.

I unlock the keeper-door and feel the kick of adrenalin as we step into the den with Komi.

He weighs 110 kilos at this point. I weigh about 70, as a comparison.

Scott hands me one side of the stretcher and we lay it on the floor. Twilight Scott is a lion keeper, calm and unshakeable, exactly the sort of person you want with you when anything dangerous happens.

Michelle directs and on the count of three we all heave Cheery Komi onto the stretcher. His arms and legs flop and he sighs.

'I got eye movement,' Hannah warns us.

More vets queue up in the corridor and gather around to intubate him, a process where they insert a long orange tube down his throat to sit in his trachea, enabling us to keep him anaesthetized and breathing properly.

I always like the more up close and personal role, being an extra pair of hands for the vets to point torches down animals' throats or hold their tongues out of the way. It gives the most intimate and tactile view of the animal and there is a certain amount of additional husbandry that can be done when they're unconscious; trim fingernails, get wax out of their ears, take close-up photos and administer vaccinations.

Komi gets intubated, the vets sliding a silver steel guide tool down his throat to enable the careful insertion of the tube. He coughs and gags as we lift him up onto the shelf on his stretcher. I climb up after him and get in position, lean my back against the 45-degree angle mesh wall.

We heave Komi up, so I can straddle his shoulders, a leg on either side of his neck with his head leaning back against my stomach.

From outside in the keeper-corridor, the vets thread two corrugated plastic breathing tubes through the mesh, connected to the anaesthetic machine.

Teresa attaches them to the tube down his throat and I hold the lot in place.

Michelle is already at work, and I have a spare hand to apply pressure to bring up the veins in his arm.

The vets are quick and efficient, with the occasional hold-up when the batteries run out in the ophthalmoscope, the thing you use to peer into a patient's eyes.

Once Michelle has done his blood tests, I use my free hand to hold his eyelids tight, first the left then the right, stretching his skin taut so Teresa can insert a tiny needle to check him for TB. Avian and bovine, one in each eyelid.

The pre-export checks are ticked off once completed, methodical, practised and planned. The whole time I feel Komi's huge bulk lying against me. His head is heavy, hair flecked with dandruff across his scalp. His arms, twice as thick as my own, slump from his huge shoulders.

Hannah crouches next to Teresa, her blue-gloved fingers in his massive hand, feeling for any twitch or sign that he is waking up.

Teresa hands me the chip scanner and I tilt his head forward, so I can run it across his broad shoulders. Wait for the beep as it confirms his microchip number.

Coughs sound from the other gorillas.

We all pause.

The group are shut away in dens one and two. They can't see what we're doing through the covered drop slides, but they know something is up and are all on edge. The coughs escalate and break into ear-piercing shrieks and screams. We hear the tumble of gorilla feet and Bob's voice away downstairs in the keeper-corridor.

The vets go back to work as the screaming desists and we reach the point of taking blood for Kera, when Komi suddenly sits up.

No warning at all, no eye movement or twitching digits.

Relaxed and heavy against me, his shoulders and neck flood with sudden strength. He lifts a clumsy arm and sits up.

In that moment, I feel myself go light in preparation for instinctive flight. My whole body ready to leap over the top of Komi's head and swing out of the den through the keeper-door, I feel like I've almost already done it, so primed am I to escape.

'Easy, big fella,' Teresa says.

Without pause Michelle produces a syringe from her pocket, flicks off the cap and injects Komi in the shoulder.

'All okay?' Michelle asks, her tone a controlled calm, and the vet team performs some wizardry with knobs and dials to knock Komi back out again. The power and raw strength begin to ebb away and he slowly leans back against me.

I wonder if Michelle and Teresa's hearts are pounding like mine.

We get the blood we need for Kera. The checks are all completed, and the den is emptied of equipment and people. The tube is removed from his throat last of all and both Teresa and Hannah remain in the keeper-corridor to wait for him to come round.

Everyone else leaves and I head downstairs to see Bob and Kera. I need to do another supervised bottle feed with Afia before I take her home tonight.

I join them in the isolation den. Kera is asleep on her foam mattress. We've got to starve her overnight, in preparation for tomorrow's knockdown, so there's no smorgasbord of foodstuffs laid out in futile attempts to encourage her to eat.

'How did it go?' Bob is lying on a pile of vet bed next to Kera on the other side of the dividing mesh. The vet bed we use at the zoo comes on a big roll, so it can be cut to size; grey, thick wool on one side with a plastic woven backing. Afia is huddled in the crook of his arm, her face on his chest, fast asleep.

'Komi woke up halfway through, but apart from that it went fine.'

'He didn't, did he?'

'The anaesthetist knocked him straight back out again,' I say. 'All this sudden strength in him, got the adrenalin going.'

'Bet it did.' He yawns. 'You ready to feed the munchkin?'

'Yep.'

'Let her wake up a bit first,' he says. 'Feed's not due for another ten minutes.'

'How's Kera been?'

'Took her meds, thank fuck.' He sits cross-legged on the floor and shifts Afia into his lap.

'What are you doing tonight?' I say. 'Sleeping?'

'Maybe. I don't know. I've been seeing someone.' He rubs a hand through his hair and pulls out some wood-wool.

I sit on the den floor opposite him and look through the mesh at Kera. She's motionless, sprawled on the foam mattress with her back to us, the sheet rucked up around her feet.

'Is this the one who works for the BBC Wildlife Unit?' I try and remember her name.

He nods. 'Yeah, but she's got the hump with me.'

'Why's that?'

Afia snuffles and her cheeks wobble, but her eyes remain closed.

'She wants to meet the munchkin of course,' he says.

'Oh right.'

'I've told her, we got to be careful these first few months, can't have her picking up a cold or something off one of us. Poor little mite got to get her immune system going first.' He yawns again. 'She understands,' he says, 'but you know what she's like, obsessed with bloody gorillas.'

I know he's reminding me of the hand-rearing protocol without telling me not to let my family get too close when I take Afia home for the first time.

He sighs. 'Thing is, Al, I really like her, but all her work mates get right on my nerves. I can't fit in with them, going on about digital animation.'

'Don't worry about them, mate, they're not hand-rearing a gorilla, are they?'

'I'm like Kera whenever I meet them. Ain't got a clue what they're on about. Puts me on edge.'

Afia opens her eyes, blinks and yawns.

Bob checks his watch.

'I'll get the bottle ready,' I say. 'I'll do everything, jump in if I'm not doing it properly.'

'Righto.' He gets to his feet. We leave the sleeping Kera and head upstairs to the messroom. The tables are still covered in vet kit for her procedure tomorrow morning.

'Hi,' Hannah calls. She's around the corner at the kettle. I can hear it bubbling away.

We join her and Bob slumps into a chair, Afia across his chest.

'Komi's up,' Hannah says.

'Does he look cheery?' Bob asks.

'We're leaving him shut in dens eight and nine overnight,' she says. 'Still a bit wobbly.'

'Great.' I pull on disposable gloves.

Hannah takes the sterilizing kit out of the microwave.

I pick up the milk powder and double check the back of the box. 'Thirty mil is one scoop?' I ask.

Bob nods from the chair. 'Sure is.'

'You heard Komi woke up halfway through?' Hannah gives me the shared wired look when you've both been through something dangerous. 'I didn't feel a thing,' she says. 'No movement at all in his fingers.'

'It felt like he inflated,' I add and reach for a plastic baby bottle. 'Shoulders and back, biceps, his chest, all just coming to life.'

'Michelle and Teresa didn't even flinch.' Hannah passes me Afia's bottle, already made up and ready to go.

Bob shakes his head. 'Let Al do it,' he says. 'He wants the practice before tonight.'

Hannah winces. 'I've sterilized all your bottles for tonight, so all they need is boiling water.'

'Thanks,' I say.

She smiles and leaves me to it, sits on her hands next to Bob at the table. 'What do you think about training Romina to be a surrogate, instead of trying to get Afia back with Kera?'

The Higher-ups had discussed options with us yesterday, but we haven't had a chance to talk about it since.

'It's such a shame for Kera,' I say.

Hannah rocks on her chair. 'She'll miss out on being a mum.'

Bob nods. 'Yeah, I'm gutted for her.' He strokes Afia's tiny back with his big hand. 'But we've got to do what's best for this little one too.'

None of us are quite willing yet to engage with the prospect that if this last-ditch attempt, Kera's knockdown tomorrow morning, fails, she'll no longer be with us.

The kettle coughs and splutters.

We'd tried to encourage interactions between Afia and Kera, but the initial interest had waned and Kera was so ill and flat she barely had enough energy to sit up.

The hand-rearing plan was to encourage interactions between Afia and the rest of the group whenever possible and Romina was already very keen on her – not quite drain sludge interested and yet not far off.

I line up the four sterilized feed bottles. 'Did Romina come and see Afia today?' I ask him.

Bob laughs. 'She was down at den two window the whole time you were up with Komi. Grumbling, stroking the munchkin's back through the mesh, giving me clumps of shitty wood-wool.'

Hannah, still sitting on her hands, keeps an eye on what I'm doing. 'Jock loves her,' she adds.

Bob nods. 'Romina as a surrogate is the safest option for this one.'

Afia wobbles her head about.

The kettle switches itself off.

Hannah's hand appears on the table, twitches like she's going to step in and pour the water into the bottles.

'Afia touched her through the mesh today,' Bob says. 'Romina was trying to lick her fingers. Gave me a bloody heart attack, seeing them teeth so close to her little nails.'

I pour boiling water into each of the four bottles, allow a little extra for evaporation and check the teats are all the correct size, number ones. 'Did anyone else come over today?' I ask Bob.

'Kukena reached through the mesh and tried to grab her arm. *Moki Mark two* I'm going to start calling her, if she carries on. Romina had a pop at her.'

'That's another good sign from Romina.' I squeeze the first gel teat into the screw bottle cap.

'Then Jock come barging in and gave Sal a clout of course. Trim Touni backed her up and they had a right old set-to.'

'I wondered what all the screaming was about.' Hannah stands up, steps to the sink and pours some water into the jug to warm Afia's milk.

'This one shat all down me,' Bob says. 'Most of it went in me pocket, all over me keys. Wouldn't have been so bad if Touni hadn't got involved. She's well cranky, thinks she's top female now.'

I sit the bottle of milk in the jug to warm.

Hannah nods. 'Touni's still tiny.'

'Got straight in the thick of it,' Bob says. 'Jock didn't know if he was coming or going.'

I get the rest of the teats ready, seal each one with a cap lid.

'Hopefully she'll get pregnant,' I say. 'Have you seen any matings yet?'

Bob shakes his head. 'Not yet, but we could do with a little playmate for you, Munchkin.'

Afia lets out a soft cry.

'None of that,' Bob coos. 'Got to warm your milky up.' He rubs her back. 'Don't want your tum tum to hurt, do we?'

I try the milk but it's still too cold.

His phone pings in his pocket.

'Do you want me to take her?' Hannah holds out her arms.

Bob passes Afia to her.

Afia wrinkles up her face and her eyes dart around the room.

'Who is this now?' Bob mutters as he fishes his phone out. I see him frown as he reads the message.

Hannah rocks Afia. 'It's alright, darling. It'll be ready in a minute. Yes it will.'

I double check the lids carefully. Can't have a leak in the hand-rearing bag on the way home.

Bob mutters and scowls at the message on his phone.

Hannah steps over, tests the milk on her wrist and Afia almost throws herself off her as she waves both arms for the bottle.

'Milk's ready,' she says.

Bob pockets his phone. 'Right.' He turns to me. 'Ready to do the feed? Hannah, hand her over.'

Chapter 8

Home at Last

I follow an ethical code as a zookeeper and strike the balance between building a working relationship with the animals, while encouraging them not to imprint on me. The idea is the complete opposite to having a pet; zookeeping relies on instilling natural behaviours, as part of the animals' fundamental welfare rights. In primates, social companionship is essential for their wellbeing and should be comparable to how they'd live in the wild. The gorillas' social interactions with us as keepers occur at a clearly defined point in the daily routine and one-to-one contact is limited to training sessions and in Kera's case play behaviour, to relieve the stress she is suffering from missing out on the company of other gorillas. The gorillas spend the majority of their time alone, without keeper interaction or members of the public about, so they have to be able to get on without us.

The method of hand-rearing Afia feels completely at odds with this approach. It has been trialled a couple of times in Europe with mixed results and has never been tried in the UK.

Rather than be hand-reared like her mum Kera until she is old enough to fend for herself, instead, a keeper will be with Afia 24 hours a day. We, the hand-rearing team, will act as a gorilla parent, while working to reintroduce her into the group before she is a year old, either with Kera, or Romina as a surrogate mum. There is no on-site accommodation at the zoo, so 24 hours a day means Afia comes back with us from work, into our homes and sleeps in our beds.

Keeping primates as pets is a huge problem. Cute baby monkeys and apes get ripped out of the wild and sold as part of the illegal pet trade. Most die of malnutrition, but any primates that become adults, driven by their instinctive behaviour, become aggressive and dominant around food. They are then generally confined to a tiny space and live alone, condemned to a life of solitary confinement, an animal like us, that thrives in company.

None of us as keepers would support having a pet primate. However, I am unprepared for the instinctive eruption of parental affection Afia awakes in me.

She is tiny, skeletal, with mad wispy hair on top of her semi-bald head. Her hands and feet look enormous at the end of her long, wiry limbs. Her nose remains flat and underdeveloped beneath curled eyebrows and wrinkled grey skin.

We dress her in a tiny pink hoody when we take her in the car seat, to keep her warm in the chill wintery air. She's asleep but instinctively grips the car seat straps with her long fingers, eyes rolling beneath closed lids as she dreams of who knows what.

The drive home is easy and with some miraculous stroke of luck, there's a parking space right outside my front door. (I live on a street lined with Victorian terraces.)

There is no way Lorna will miss our arrival. Our neighbour over the road. The lovely Lorna, who takes in parcels for us and has a spare set of keys. She is up and about when I leave for work in the morning, crisscrossing the road with papers or bottles of milk. When the kids were younger, Lorna would watch out for them coming home from school and let them in when they lost their keys. She knows everyone on the street and is truly lovely.

I scurry around to the passenger side and unclip the car seat.

Lorna waves at me through her front window and points with a big *Oh* look on her face.

I lift the car seat out and her door opens. She beetles over, trailing a long blue cardigan and wearing glasses with a bead string. She always smells the same.

'Look at her,' Lorna whispers. 'She looks like a little ET, doesn't she?'

Afia's eyes open, her view half obscured by the pink hoody. She lifts one of her spidery hands.

'Look at those little nails. She's gorgeous. Takes after you, does she?'

'Her eyebrows aren't far off.'

Lorna cackles.

'We've got to keep it secret that I'm bringing her home,' I say.

She nods. 'I'll let Janet know. Between us we'll keep it quiet then, won't we?' she says to Afia.

'She can't be exposed to lots of people yet, she's premature so super vulnerable.'

'Bless her,' Lorna says.

My partner Sparky opens the door and I ferry Afia inside and down the hall. Plonk the car seat down on the kitchen table.

Sparky has always been supportive of my job. She lives with the reality of me being at work every other weekend, missing at Christmas and Easter, coming home hours late after animal clinics and getting up ludicrously early. She's interested in the gorillas, knows who they all are second hand as I never cease banging on about them, but she's never been up close to one before.

'Hello.' Sparky offers Afia a finger to hold. 'I've washed my hands,' she adds.

We've agreed that our partners can have limited contact, the thought being we would probably already be sharing anything contagious.

'I've just got to grab the hand-rearing bag out the car.'

Lorna is on Janet's doorstep two doors down. She gives me a thumbs up as I heave the hand-rearing bag out of the car.

Afia is half awake and gazes around the room. Her head moves in uncoordinated jerks, at constant risk of over balancing. I unclip her from the car seat and Sparky helps me disentangle her arms as I pull off the hoody and Afia's fingers and toes grip my string vest as I put her on my hip.

'Does she have to be on you the whole time?'

'If she was in the group,' I say, 'she'd be clinging onto her mum, mainly attached to their wrist, but part of her development is to learn to ride on their backs.' I take her teething chew out of the hand-rearing bag and we all go into the lounge.

'I didn't realize how tiny she was going to be,' Sparky says as she sits with us on the couch. 'I've tidied everything away. When's her next feed?'

'When she properly wakes up,' I say. 'Then once every two hours.'

Afia pushes herself up off my chest, arches her knobbled spine and her lips open, tongue rolls and a massive yawn seems to take her by surprise. Mouth wide, bright pink shiny gums, nostrils flared, her hair in a quiff above her eyebrows. Eyes screwed shut. Her head wobbles and she thumps back on my chest, shuffles and grips the vest. Back to sleep again.

'Sam is giving us a call later,' Sparky says. 'And Fizz is staying with their dad this week.'

'Is Sam going to Facetime?'

She nods and beams down at Afia. 'Hello you,' Sparky says in the tone of voice I've heard her use with puppies she finds particularly cute in the street. 'Are you waking up?'

Afia raises her head again and blinks her eyes, tries to make sense of the wall behind the sofa.

Sparky is amazing. She steps into action with the milk and brings it all into the lounge.

The feed is terrifying. Afia is figuring out how to suck from the bottle as much as I am learning how to feed her. I'm desperate not to let her aspirate – to get milk on her lungs – so I keep the feed slow, count each swallow and take the bottle away to pat her bony back with two fingers, to wind her.

'We need to get a certain amount into her at each feed, so she'll put on weight.'

Afia releases the gurgling burp.

'How do you know how much she should weigh?'

'It's in the guidelines we're following,' I reply. 'A plan that follows each stage of her development, so she keeps pace with a baby gorilla of the same age being reared naturally.'

I offer Afia the bottle again and count the number of sucks.

'She's at the eat, sleep, repeat stage,' Sparky says. 'I remember it with the kids.'

I got together with Sparky when Sam was eight years old and Fizz four. I'd never experienced the looking-after-babies part of parenthood.

Afia falls back to sleep after the feed and Sparky sets up for a Facetime call with Sam.

Sam's face appears on the screen and I see surprise followed by awe and wonder. 'No way, is that... wait, it looks like a little gremlin.'

'Our newest resident,' Sparky says.

'Is it sucking its thumb?'

We've not seen Sam since Christmas, when he went back to university.

'She's asleep,' I say.

'That's so insane. You're sat with a gorilla on the sofa.'

'How are you doing?' Sparky asks. 'How's the course?'

'Fine,' Sam says. 'What do you do when she wakes up?'

'I've got to be as gorilla-like as possible.'

Sam laughs.

'We've got to prepare her for going back into the group,' I explain. 'Once she gets to an age where she will come to us at the mesh for a bottle feed, we'll be introducing her to Kera or possibly Romina as a surrogate mum.'

I can see Sam is vague as to who Kera and Romina are. 'How long will that take? Years?'

'If we get her back with the group before she's a year old,' I say, 'she won't remember any of this, the sofa, or us. She'll grow up as a normal gorilla.'

'So the plan is to give her back to her mum, then?' Sam asks.

'If we can. Kera's super sick. The vets are going to try and save her life tomorrow.'

'Her mum is the one that never fits in,' Sparky adds. 'Ostracized by everyone else.'

Sam nods. 'Right, yeah I remember now. Shit Al, that's going to be intense.'

'Hopefully, it'll go okay,' I say to Sam, Sparky and Afia. 'If we get this hand-rearing right, there'll be no negative impact on her later. Don't want to mess your kids up, right?'

Sam laughs.

'Fizz could do with a call,' Sparky says to Sam. 'While I think of it. They're in GCSE revision misery.'

'Our other member of the family troop,' Sam says.

Whereas I had missed out on baby parenting, Sparky of course had not.

'She's got colic,' she says to me after the next feed.

Afia whimpers and calls as I rub her back.

'I thought that was a horse thing?'

Sparky shakes her head. 'Look at the way she keeps pulling her knees up.'

I walk about with Afia, show her the house plants, books on the shelves. We wander out into the kitchen and back into the lounge and eventually she calms down. I sit with her and she lies in the crook of my arm. Her feet still jerk at random, but she's not whimpering and grabs her ankle as the sole of one foot pokes her chin. She shuffles and her body shudders as she shits liquid gunk into her nappy. I lie her down on the changing mat and she shivers at its touch.

'Sorry, littlun,' I say.

The nappies she wears are the tiniest available, for premature human babies, but they're huge on Afia's scrawny body. She kicks her legs as I clean her arse with cotton wool and warm water. New nappy on and I head upstairs for the first round of sleep in our bedroom.

I lie Afia in bed while I strip off my clothes. I've never been one for pyjamas and as I pull the string vest back on, I catch sight of myself in the mirror. Naked apart from a string vest is not a great look for me. I pull on some pants and climb into bed, lie on my back and lift Afia up onto my chest. An infant gorilla would sleep on its mum all night. She grips the string vest with arms and feet and I know that I need sleep and will have to be awake in an hour and a half for a feed, but there's no way in hell I'm nodding off before then. What if she gets too hot? I tug the duvet down and rest my hands across her back.

She snuffles and burps a cloud of cheesy milk breath. Her breathing rasps in and out and flicks between strings of quick pants and sighs. I feel her heart on my chest and her grip relaxes.

I think of Kera lying on her foam mattress at the zoo. Will she be alive this time tomorrow?

My alarm buzzes. I must have fallen asleep after all. Midnight. Afia bumps her head under my chin. The soft fur on top of her head brushes my nose. She is rooting around my neck and dribbles.

The bedroom door opens and Sparky appears with the warmed milk. 'How's it going?'

'You legend.' I sit on the edge of the bed.

'That's an interesting look.' She hands me the bottle.

'She keeps making all these snuffly noises, breathing's all over the place too.'

Sparky smiles. 'The kids would do that when they were tiny, don't worry about it.'

Afia reaches for the bottle.

'I'm going to bed,' Sparky says. 'Good luck with the rest of the feeds and call me if you need to.' She crouches down. 'Good night you,' she says to Afia.

Afia takes the teat, eyes drowsy and head loose. I keep the bottle tilted and sit her up. Her eyes spring open and the lids droop a second later. She slowly sucks. The milk makes her heavy with sleep. She takes it all fine and I pat her back until she does a massive body-quaking burp. We both snuggle back down to sleep again and her body relaxes across my chest.

The alarm buzzes and I think I must have pressed snooze, for it to go off again so soon. But no. It's 2 a.m.

I change her nappy first and take her with me as I go downstairs to warm the milk, keeping her pressed against me in the crook of my arm. She stays asleep as I boil the kettle and head back upstairs to bed with the milk in a tub of hot water. I sit her back in bed and shake the bottle over my wrist to check the temperature. Still too cold. I realize I haven't filled in the clipboard for the 12.00 feed yet and dig about in the hand-rearing bag for a pen.

Afia's whole head is hijacked by another wobbly massive yawn.

I check the milk and it's ready, sit her up and offer her the teat. She clamps her mouth around it and takes most of the bottle. I tap her back in between sucks. Ten mil left to go and we can sink back to sleep again.

She hiccups.

A dribble of milk runs out of her nostril.

'Shit.' I pat her back. 'Fucking balls. I'm sorry, darling.' I hold her upside down and she clutches my arm with both hands. Hiccups again and more milk dribbles onto her top lip.

I've aspirated her. First night and I'm fucking it up. And I left the nebulizer out in the car.

She squirms and her foot clutches my arm. I sit her back on my knee. Tap her back. Her head wobbles on her neck, her eyelids droop and she burps. I tuck her against my chest and gently continue patting her back, pick up the pen and scrawl on the hand-rearing clipboard:

Feed: *2.00 AM*. Amount: *70 ml*. Mood: *Sleepy*. Urine: *Yes*. Faeces: *No*. Notes: *Aspirated. Milk from both nostrils. Held upside down and winded*.

She snuffles and I think about calling the vets, but I know I'm probably overreacting, vow to be slower with the next feed. I got overly confident as the first few had passed off without incident. She's still getting nebulized twice per day anyway, but I've not made things any better and then the 4.00 a.m. alarm goes off.

Sparky appears at the 6.00 a.m. feed with the milk made up and a cup of coffee.

I take Afia downstairs for the 8.00 a.m. bottle and we huddle around the kitchen table.

'How does her breathing sound?' Sparky asks after I fill her in on the aspirating feed.

'There's no raspy noises.'

Sparky pulls her chair close to listen and I rub Afia's back. The skin around her delicate ribs is baggy, like a suit she'll grow into. She blinks and jerks her head to take in the bookshelves and reaches a hand for Sparky.

'I'm going to get going,' I say. 'Take her into work and nebulize her.'

'You what her?' Sparky asks.

'The nebulizer, it turns medication into mist and she inhales it through a mask, fights off the threat of pneumonia.'

A key sounds in the lock and we hear the front door open.

'Hello.' It's Fizz. 'Just come to grab my chemistry book.'

'Hi, darling,' Sparky says. 'Look who's here.'

Fizz pauses.

I have Afia in my arms and she has her tiny hand around Sparky's finger.

'Why is she bald?' Fizz asks.

'She's got some hair,' Sparky says.

Fizz laughs. 'Her body, I mean. That hairstyle is crazy, like my troll doll, remember? Sam threw it out of the window when we went camping.'

'Are you a troll doll?' Sparky asks Afia, in her talking-to-a-baby tone. 'Are you?'

Afia turns her head to gaze at Fizz.

'Oh, she's so cute,' Fizz says. 'Honestly.'

Sparky helps pack Afia's bag as Fizz searches out the chemistry book upstairs.

'I might be late tonight,' I tell her. 'Depends how we get on with Kera.'

Sparky squeezes my arm. 'Good luck,' she replies.

Chapter 9

Last Chance Saloon

Afia mainly stays asleep as I nebulize her.

She wakes up just at the end of the allotted 10 minutes and reaches for the mask and tubes, but mostly she stays still on my lap. I flip the off switch and fill out the clipboard. Afia grips my vest as I take her downstairs to see Kera.

Kera lies on her foam mattress, as I imagined her last night, flat and almost lifeless.

Bob has emptied the front half of the isolation den and scrubbed the floor, got everything as clean as he can before the vets arrive.

'How was last night?' he asks.

'Not too bad,' I say. 'Aspirated her once at two in the morning.'

He nods. 'We're doing our best, mate, don't worry about it.'

Afia wobbles her head around to look at him.

'Hello, Munchkin,' he says and tickles her neck.

'What time are the vets coming?'

'Ten allegedly, but we've got some scanning specialist coming in, so whenever they get here.'

'Does she snuffle all night long with you?' I ask.

He nods. 'Makes you think she's suffocating, don't it?'

I yawn.

Kera lifts her head from the foam mattress and grumbles. She is curled up with her feet pressed against the mesh. Even her tone is tired; the grumble comes out as a low murmur.

I take Afia over to see her. 'Who's this, gorgeous?' I say to Kera.

'Fingers crossed for today,' Bob says behind me.

My eyes tickle with sudden tears at the thought of Kera not being here tomorrow.

'Least the vets can get all her meds onboard,' he adds. 'Spare us that today at least.'

Afia waves one of her spidery hands and I help her touch the sole of Kera's foot through the mesh.

'Once Kera's under, you're going to lie Afia on her again,' Bob says.

'Great idea.' The rota means I'll be looking after Afia all day today, so I won't be involved in the vet procedure with Kera until she is under anaesthetic. We'll place Afia on Kera when we can, so she can feel her mother's fur and warmth.

Afia has limited, jerky control of her hand, as if she's being controlled by puppet strings. She pats at Kera's foot through the mesh.

Kera grumbles again.

'See if we can reawaken anything, give Kera a lift if nothing else,' Bob says.

He's got an intense air about him today; I know the frown and tense shoulders are a sign of his internal battle. He's

Photo of Alan and Hasani on page 3 © Imogen Calendar; all other photos © Miriam Haas

Kera, keeping out of the way as usual

Afia, 15 months old

Romina shares a sliver of her favourite food, beetroot, with Afia (1 year old)

Afia clinging to Romina's arm for comfort

Hasani looking very weak on Kala

Afia breaking things

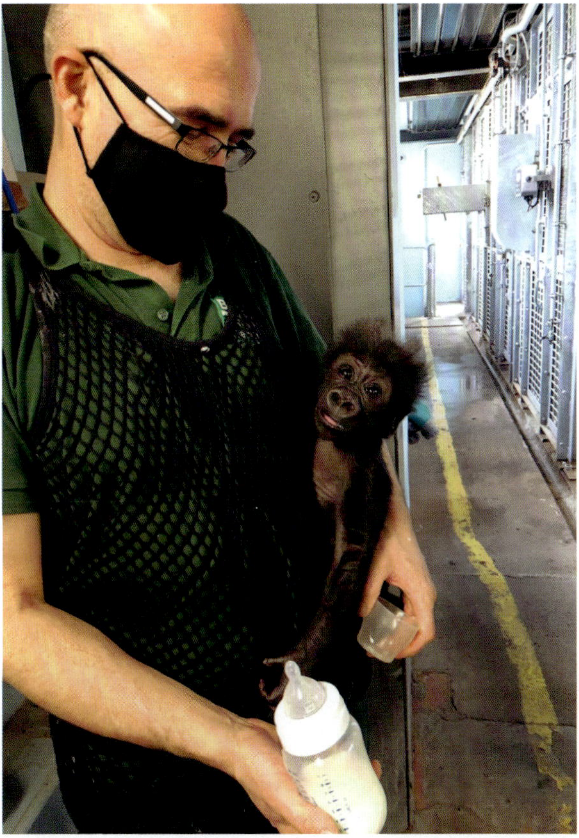

Alan bottle-feeding baby Hasani

Afia, 6 years old

Hasani (10 months old)
exploring the island

Kera, 18 years old

Kala with her log

Perfect Juni (18 months old) riding Trim Touni

Jock with his family

Perfect Juni hoping for some iceberg lettuce from his dad, Jock

Romina keeping an eye on Afia

Dangerous Sal and Kukena, plotting

Ayana learning to cling to her mum Touni's arm

*Afia figuring out what's
edible on the island*

*Hasani at 1 day old
with his mum, Kala*

Kala clapping her chest

Juni being perfect as usual

Afia destroying her bed sheet

Hasani looking nervy

Jock relaxed in the sun

*Hasani discovers the
delights of yellow pepper*

Another perfect morning for Juni

refusing to think about what it will mean if we fail and has taken the step into immediate response mode; he's ready, like all of us, to throw everything at Kera today to save her life.

'Oh yeah,' he continues. 'When you do walking with the munchkin later, watch out she don't eat no wood-wool.'

'Good tip.'

'She stuffed a handful in her gob yesterday and was rolling about with it.' He crouches next to me. 'Bit unsteady on the old pins.' He tickles Afia's neck. 'Aren't you?'

'I need to do a milk feed,' I say.

'I'll come give you a hand. Might be last chance for coffee before it all kicks off an' all.'

'See you later, Kera,' I say, but she doesn't respond, asleep again on her foam mattress.

'Love you,' Bob adds.

We have the messroom to ourselves as it's still early.

'How's her hanging on going?' Bob asks.

'Pretty good.' I turn and show Bob Afia. Her hands clutch my vest, but her feet flop off from time to time. I support her with the crook of my arm but keep it as light as possible.

'Look at you go, Munchkin.' Bob does his coo tone. 'Getting there.'

I turn on the kettle.

'I'll make up the bottle,' he tells me. 'You look like you've been up all night.'

Afia bounces her head against my chest as I sit down.

Bob clatters about by the sink. 'Here,' he says. 'What do you think of Hannah?'

I yawn. 'How do you mean?'

He unscrews one of Afia's bottles. 'Is it just me, or has she got, like, you know... hotter?'

'Hannah?' She's like a sister to me at this point, much as Bob feels like my brother.

'Something about her.' He turns to the coffee machine. 'Not noticed it before.'

Afia bumps her head on my chest. Mouth open. Her body, driven by hunger, is rooting for a gorilla teat.

'How did it go with...' I can't remember the name of Bob's girlfriend. '...the BBC one?'

Bob thumps the filter with his palm and yesterday's cold ground coffee thuds into the bin. 'I'm just knackered,' he says. 'Ain't got time for relationships. I'm in bed by nine.'

We hear the door open downstairs, and both recognize Hannah's footsteps and the chink of her keys.

'Here she comes,' Bob mutters.

Hannah trots up the stairs. Zipped up in a fleece, with a woolly beanie hat on. 'Morning.' She waves at Afia. 'How's Kera?'

'Alive,' Bob replies. 'Coffee?'

'And how's my little munchkin?' Hannah says in a talking-to-a-baby tone.

Afia wobbles her head and cries.

'Darling, are you hungry? Is Al starving you?' Hannah clucks. 'How did you get on last night?' she asks me.

'Aspirated her at two this morning.'

Hannah glances at the nebulizer. 'It's hard,' she says. 'Until she learns how to moderate the suckle response.'

Bob pours a hiss of fresh coffee into the machine. 'This'll sort us out,' he says.

'Did you get any sleep?' she asks me.

'Yeah. An hour or two here and there.'

Hannah smiles. 'It'll get easier. Do you want me to feed her?'

I choose not to be offended by the slight tone of criticism, accept the help instead. 'Legend.' I hand Afia over. 'I need to rewrite last night's notes, make them legible.'

Hannah holds Afia on her lap, checks the milk temperature on her wrist and lets her suckle.

Afia slaps her hands at the bottle and rocks as she sucks.

'Steady,' Hannah says softly, slowing the feed as Afia's cheeks billow in and out.

I look around at Bob and he gazes at Hannah and Afia.

His mouth is half open, lips ready to smile, eyes wide and stone still. 'Look at you two,' he whispers.

The messroom fills up, the rest of the keeper team and the vets again in a muted repeat from yesterday. Plans laid out and risk assessment signed. Wishes of good luck as those not involved file out to work across the other sections.

I wait with Afia in the keeper-corridor downstairs. The far end is swept clean and free of veterinary equipment. I sit near the mesh window overlooking den two and Jock's training platform. We've booted the gorillas out for breakfast, but it's cold, so once Kera is out and the vet equipment in, I'll give them all access back into den two. My role in the procedure is Afia and keeping the rest of the group calm.

I sit on the corridor floor. One of Afia's developmental goals this week is walking and the floor of the keeper-corridor is smooth concrete. We start on the torn mat of wool-topped vet bed.

I sit Afia on the mat next to me and the woolly pile provides gorilla fur-type clutching opportunities.

'Ready?' I say and gorilla grumble to reassure her.

She clutches the woolly vet bed with her fingers and toes and pulls herself off the floor.

'You're doing it,' I tell her.

She wobbles and shakes, stands up on all fours, puffs her cheeks in and out, not sure if she should hoot in panic. There is no coordinated control of when she releases her tight grip on the vet bed and she face-plants, chooses to roll onto her back as she is already halfway over and lies there, eyes fixed on the ceiling. She flails an arm until she catches a kicking ankle and shoves her toes in her mouth almost accidentally.

This is where the smooth floor comes in handy. I select a clump of wood-wool, remember Bob's warning about her trying to eat it yesterday and pick up Afia. We shuffle down the corridor and I place the wood-wool in a small pile on the concrete floor. The underfloor heating down here makes the concrete warm and none of the gorilla shit is hosed up this end, so it's dry and clean. I crouch and put her on the opposite side of the wood-wool, so the clump rests between us. Place her hands in it, facing me, and back away from her, to sit cross-legged on the vet bed.

Her lips purse and wobble in fear and I fight the urge to sweep her up and comfort her, grumbling at her instead. She struggles to her feet, toes clenched into balls on the bare concrete floor, fingers entwined in the wood-wool. Her stick legs jerk and force the wood-wool mat to slide forward even as her arms remain still.

'You're doing it.' I grumble encouragement as the wood-wool acts as a gorilla baby walker.

Down the corridor Bob hand-injects Kera. The mesh slides that look out of the isolation den are all covered up with tattered duvet covers. The vets begin to bring in the kit they'll need to give Kera the blood transfusion.

I listen to their hushed tones and the clatter and clang of vet equipment as Afia begins to push her clump of wood-wool repeatedly up the corridor.

Once the equipment is in, I give the gorillas access into den two.

They prowl and stalk about on the upper level, taking it in turns to peer through the slides and mesh into den one at the curtains that block the isolation den and Kera from view.

Eventually Romina appears on Jock's training shelf and grumbles at me.

I sweep Afia up and sit on the wood-wool bales, opposite Romina on the other side of the window.

Romina wedges her fingers through the mesh.

Afia stares and I press her tiny hand on top of Romina's fingers.

Romina exudes her pungent cut-grass smell. She shifts her weight on the training platform to press her belly against the mesh.

Afia wobbles on my knee. She clutches the skin of my arm with one hand and pats Romina with the other, or perhaps her hand is flailing uncontrollably – it's hard to tell.

I support her back and we edge closer.

Romina shifts about. Leans forward on her elbows and presses her nose to the mesh.

I hold Afia's hand up for her to sniff.

She grunts.

Trim Touni materializes, taking time off from parading about as Jock's new favoured female, having climbed her way to the top of the hierarchy. She stands upright on her legs and cranes her neck to catch a glimpse of Afia.

Romina's flat pink tongue wraps around Afia's fingers.

She pulls her hand away.

Romina stuffs a wedge of folded cardboard through the mesh, eyes fixed and intent, as Kukena's head appears, upside down above the keeper-window as she leans down from the upper level. Dangerous Sal's daughter Kukena has got the classic I'm-going-to-annoy-you look on her face and sways, a rope clutched in one hand.

Romina coughs, three quiet grunts.

'Good one, Romina,' I say.

Kukena's head disappears again, but I don't hear the shuffle of feet, so I know she's still above us.

Romina controlled the volume of her cough, so neither Jock nor Sal would hear her low warning to Kukena.

Afia reaches for Romina's nose.

A huge pile of wood-wool falls onto Romina's head, shoved off the edge by Kukena.

Romina coughs and stands up on the training platform, swipes for Kukena, who scampers off toward her mum, on top of the glass hide on the other side of the enclosure.

'You tell her, Romina,' I say.

Sal glances up from her picnic of breakfast leftovers, collected from the island outside and carried back indoors to eat away from everyone else. She doesn't react.

Jock glowers and his huge chest resonates with annoyance, but he stays where he is, at the slide into den one, and keeps watch.

Kukena stalks back around to her spot above us and the keeper-window, out of sight, and carries a length of browse, the branch stripped of bark and leaves.

I look at Romina. We both know Kukena is going to use the stick to poke her head.

I hear footsteps in the corridor and see Hannah, face covered with a surgical mask. She gives me a thumbs up.

I grumble at Romina and take Afia away down the corridor. Hannah holds her as I fit my own mask and we go into the isolation den.

Kera is flat on her back, her head supported by a pillow. The tubes are inserted in her mouth, held open by strings of crepe bandage in her teeth. She lies on piles of vet bed and has the electric blanket on top of her.

The atmosphere feels pretty good, I can see it in the way the team all move.

'Stitches have healed incredibly well,' Michelle the vet says from behind her mask.

A bag of blood hangs above Kera.

The vet team make room for me, and I crouch, feel the coarse stubble on Kera's shaved belly through my gloves as I lie Afia on her chest. Hold her so she can grasp her mother's fur. We let Afia lie on Kera right back in the early days of her illness, but she was so tiny then and unaware of her surroundings.

We watch as Afia flexes her fingers and toes and consciously clutches Kera's fur for the first time.

I sit with Bob in the isolation den afterwards as we wait for Kera to come round. Her bedding is all fresh and we've positioned her on the foam mattress so that her head is near the mesh.

Hannah pokes her head in, Afia on board and the hand-rearing bag in her hand.

'How's she doing?'

'Still asleep,' Bob says over his shoulder. 'Should be coming round soon.'

'One of the vets will pop back to sit with her if either of you wants to head home,' Hannah offers.

I look at Bob and we both shake our heads.

'We'll wait a bit,' I reply.

'At least it went well,' Hannah says. 'We've done everything we can.'

'She'll be alright,' Bob states.

'I'll see you both tomorrow.' She heads off with her familiar jingle of keys.

'You didn't say nothing to Hannah, did you?' Bob asks.

'About what?'

'Saying I fancied her this morning.'

'No, mate.'

'I don't know why; I think it's something about seeing her with Afia. When she feeds her.'

'Nights in front of the fire with your hairy baby?'

He laughs.

'Not later than nine o'clock, though.'

'Bloody right,' he says.

Kera lifts her head.

'Welcome back,' Bob says to her. 'You thirsty? We got all the treats here for you.'

Chapter 10

Communicate

Kera began to recover and slowly regained her appetite after the blood transfusion and iron supplements that saved her life. Her legs were skinny, fur matted and dusty and her eyes remained droopy, but she got herself to her feet and started to eat. Not just the treats we used to get meds into her, but proper gorilla food, leafy greens by the armful, browse and her usual vegetables. We had given her body the boost it needed. She began to replenish her own red blood cells and steadily got better.

Cheery Komi left for his new home in the Netherlands, and we were back to one group of gorillas again. Seemingly none of the gorillas except for Kera missed him and it was a relief for us that he was no longer living alone. His departure meant Kera was once again friendless and the length of her illness meant she had no maternal interest in Afia. We began to train Romina daily for surrogate mum duties.

Afia is now ten weeks old and down to one feed every 4 hours, so we are all getting some sleep. The aspirating days are behind us, as she has learned how to suckle from the bottle, which means no more nebulizing, and she has started to stuff her face with solid food too.

A baby gorilla will beg from its mother, gaze up with adoring eyes, attentive and hopeful that a morsel will tumble from the churning jaws above and land within grab range. They learn to duck and dive as they shove the food into their mouths to avoid their mum's grasping fingers.

Afia's solid food starts with steamed sweet potato and crunchy leafy greens. This is the stuff we wanted onboard to get some weight on her. Instinctively she is trying to jam things down her throat and with four grasping limbs at her disposal, nothing is safe.

I pull up outside the house and put the hazard lights on. Sparky is primed and ready, and comes out of the front door as I unclip the car seat.

Over the road, Lorna shows me the usual big thumbs up through her window and I give her a wave.

Afia waves too as Sparky hoists out the car seat and disappears back inside.

I have to go off in search of a parking space and almost forget to take Clive, the toy gorilla, out of the back seat.

By the time I get back to the house Sparky and Afia are in the lounge. I can hear them playing, Sparky talking in her happy tone of voice.

'Who's got your baaaggg? What's in your bag?'

I hear rustles as Afia wrestles with said bag, her current favourite toy.

'Everything okay?' I ask.

'We wanted to get out and play.' Sparky sits on the floor with Afia. 'Didn't weeee. What's in your bag?'

The other toys are scattered across the wool rug, a selection of empty plastic bottles, a red Celebrations chocolate tin, also empty, her teething rings and the rubber spatula.

'Why have you got a cuddly toy gorilla under your arm?'

'Meet Clive,' I say. 'Clive, meet Sparky.'

'You're losing it,' she responds.

'This is Home Clive,' I say. The gorilla toy is a similar size to Afia, with silver hair on its back and a plastic face. 'She's been wrestling with Zoo Clive during the day.'

'Zoo Clive?'

'The pissy, cockroach-infested, shat-upon version,' I explain. 'Home Clive is clean.'

Afia lies on her back and pedals at the shopping bag, convincing herself it has a life of its own, as it crackles and jumps about. She grins her big white teeth at me. The black fur has grown in down her back and forearms, her armpits and chest are soft, grey, baggy skin. She flaps her arms and legs, rolls to grasp the wool rug with fingers and toes and then pulls herself up onto her feet. She seems to take herself by surprise as she veers across the rug. Her arms and legs pump at different speeds and her toes still sometimes forget to unclench, but she manages to stumble toward me, scattering the plastic bottles and chocolate tin out of her way.

'Look at her go,' Sparky says.

I crouch down and Afia clatters into me and jumps at Clive. She chews his odd plastic gorilla face and pulls him out of my arms to discard him on the rug.

I drop on all fours and sweep her up and onto my shoulders. Her toes hook onto my vest, one hand tugs my ear. 'She's started riding on our backs,' I say, 'like a real gorilla.'

'Have you?' Sparky reaches out and tickles her chin.

Afia slaps me enthusiastically on the head. 'What time's she due a feed?'

'Her next feed is her tea. Solids. Milk feed at eight.'

Afia dismounts.

'She's after the phone,' Sparky says, 'and the wifi. It's the lights I think.'

She motors across the rug in the direction of the TV and Sparky distracts her with a plastic bottle.

A key sounds in the front door and Afia freezes.

'Look at that,' Sparky says. 'She's associated the noise of the door with Fizz coming into the hall, she did it just now with you too.'

Afia looks expectantly at the lounge door.

'You're so clever,' Sparky tells her. 'Yes you are.'

Fizz appears around the door. 'I can hear you two laughing out in the street.'

'Hey you,' I say. 'All alright? How was school?'

'Revision session for mocks. Mr Harris was in a right stress about it.'

'Which one is he again?' Sparky asks.

'Theatre studies. He's freaking out.'

Afia pats a plastic bottle.

'Hi Afia,' Fizz says. 'She's getting way bigger, my god.'

Afia patters forward but gets distracted as her toes accidentally clutch the bag handle and it apparently ambushes her out of nowhere. She rolls and tumbles with it.

'You've got your baagg,' Sparky says.

I laugh.

'Who wants a tea?' Fizz asks.

'Not for me, thanks.' I get the bag to attack Afia. She lies on her back, arms and legs thrashing.

'No thanks, darling,' Sparky says.

Fizz disappears.

Once bag fighting peters out, Afia uses me as a ladder to clamber up and onto the couch and begins to methodically drag the cushions from one end to the other.

'She's started doing this as well,' I say.

Afia takes a couple of steps, braces herself with one hand against the back of the couch, hauls the cushion along and stumbles over it.

'Interior design.' Sparky looks at her watch. 'I'll get dinner.'

Afia sits on my lap with a wedge of steamed sweet potato squished between her little fingers. Her hand jabs it up her nose and smears it against her cheek as she takes her eyes off it to peer around the table at us all, her mouth open in hopeful anticipation.

My plate is way out of her reach and I time each forkful to avoid flailing arms.

'It's so weird seeing you both with a baby.' Fizz laughs. 'I'm so jealous.'

'What?' I say.

Afia jams the sweet potato in her mouth and jiggles on my knee.

'Not really jealous, but kind of. Was I like this as a baby?' Fizz asks Sparky.

'This is how I fed you,' Sparky says. 'I used to surround you with little bits and pieces, and you'd choose what you wanted – apple, then baked beans, cheese, a selection of stuff.'

Afia drops the sweet potato on the floor with a wet splat and slaps her hand in the plastic container of lettuce leaves, up-ends it and flounders for a chunk of cucumber.

'Hold on, you,' I say.

She shuffles on my lap and takes a long romaine lettuce leaf, pokes herself in the face before clamping her teeth on it.

'I wasn't as cute though, was I?' Fizz asks. 'No one can be as cute as her.'

After dinner I lie on the couch and Afia half-heartedly pulls the cushions up and down across me.

I should've spent more time thinking about what Fizz had said in hindsight. My only excuse really is how tired I am and how full of Afia my waking life has become. Fizz is right of course; we are in love with Afia. Can't help it, and despite the husbandry guidelines and developmental goals, the training we are doing with her during the day, I am also emotionally infatuated. Fizz has never seen me or Sparky with a baby. A whole host of instinctive behaviours manifest as Afia becomes part of our family group at home.

Afia abandons the cushions and clambers up me as I lie on the sofa. Her limbs wobble and I can see the concentration in her face as her toes un-clutch. She climbs onto my chest. Her mouth works with the effort, a determined jerky progress to be able to look straight into my eyes. The look is open, sentient, aware of herself and me. A willingness to understand, an expectation and what feels like encouragement, the look blows my mind. An invitation to further communication, that for us – as humans – is of course language.

Complex language is seemingly a unique adaptation that we humans picked up somewhere along the evolutionary timeline, that other species of primate did not. However, it is estimated that 70 per cent of human communication with our own species is non-verbal, and a mere seven per cent is the words we use. The rest is made up of tone, facial expression and body language, almost all of which we share with gorillas.

Afia reaches out a gentle hand for my lips.

There's a species called the gelada baboon which inhabits the highlands of Ethiopia. They are grazing primates, with thick fur coats that insulate them from the weather. To get their nutritional requirements, they have to spend about 10 hours a day stuffing handfuls of grass into their mouths, with small dexterous fingers. As a result, they have less time for physical contact, can't groom one another to reinforce their social bonds, so chatter instead. A series of mutters and murmurs that seemingly reassure the group that everything's fine.

There is verbal communication between most species of primate, to varying degrees, but complex language as we know it is said to be inhibited in every species other than humans. The other primate species collectively lack the neural control over their vocal tract muscles to produce words. Their brains never evolved a need to talk.

Afia's nose quivers and I know her sense of smell far outweighs mine. We share body language and to an extent, facial expression. She laughs like a drain when I tickle her, plays with a big smile on her face. Wobbles her lips and cries if she's scared, clings to you if shocked by loud noises, but her other senses are beyond mine, smell in particular, and so our communication will always be limited.

As a species, humans are driven to like babies. We have an instinctive protective urge that floods through us. Think of the twitch you get when you see a toddling baby about to stumble over. The speed at which you can snap out a hand to support them. There are exceptions of course, but generally in humans there is a universally shared instinct toward protecting the young. Their proximity awakens a raft of parental concerns and care.

I would do anything for my kids, a tired saying perhaps, but true, actions that don't require thinking about, but are rather an automatic response that transcends the day-to-day. In more affluent societies there is the added wave of emotion around animals too that plays to a similar protective urge. Puppies and big-eyed kittens. One of the reasons why the RNLI save dogs from drowning is to limit the risk to owners leaping in to save them. To see an animal in distress plays into our emotional drivers, not perhaps at the same level as that of a human baby, but the instinct lies within us. Afia had fired all that up in me and to a degree in Sparky. We were hooked.

Afia gives up looking into my eyes and slumps on my neck, shoves her head under my chin. She'll pass out now for an hour at least, then a milk feed, which she'll bomb at eight and sleep through until midnight.

Afia wakes me up with a series of slaps, applauding on my forehead. Sparky is asleep next to us. Now Afia is older and has built up her own natural defences, Sparky has moved back into our bedroom.

I look at the clock: 2 a.m. She's had her midnight feed and is awake 2 hours early. She has her play face on and sits up wide eyed and enthusiastic.

'No way,' I whisper and pull her back down to sleep, snuggle in the combined warmth of us all under the duvet. Her eyes droop and she wriggles under my arm to get comfy.

Slap on the head and she's up again.

'What time is it?' Sparky mutters.

'Only three,' I reply. 'What are you doing?'

Afia gnaws the headboard and drums it with one hand.

'Teething,' Sparky says.

'Come on, you.' I pull Afia onto my chest, but she shuffles around and crawls off down the bed. Tangles her feet in the duvet.

'May as well feed her,' I say, 'get her back off to sleep.'

'Have you got any Bonjela?'

Afia has reached the end of the bed.

'I'll grab it all. Are you okay with her?'

Sparky yawns. 'I'm fine.'

Afia falls out of bed. Thumps her head on the floor. There is a moment's silence then a *'who, who, who'*.

I pull her back up into bed and she snuggles in with Sparky.

The floorboards creak as I try to tiptoe past Fizz's bedroom. The lights are on and even if Fizz has fallen asleep, my clunking around downstairs and boiling the kettle will wake them up.

My alarm goes; 6 a.m. and both Afia and Sparky are asleep. It's the golden hour – the longer she stays asleep the more chance I've got to prepare everything for taking her back to the zoo. The hand-rearing bag is loaded up, notes completed, and I can go and get the car, parked three streets over. If I'm lucky, she'll stay asleep until we get to work.

I park outside the side door to Monkey Jungle. It has a locking arm inside, so I need to wait for someone to come and let me in.

Afia is awake and keen to get out of her car seat. She's not wearing her hoody, those days are gone too as she is now capable of self-regulating her body temperature and it is late spring, toward the end of April. A taste of summer when the sun was out.

My phone vibrates. A text from Hannah. *'Be with you in ten mins x'*

I help Afia out of the car seat and she clambers onto my lap. Stands and sinks her teeth into the rim of the steering wheel.

I've turned the car off, so nothing happens when she pulls and pushes on the indicator stick, but she manages to turn on the radio and is momentarily enraptured by the lights and glances over her shoulder as the morning news voices come from the speakers, to see if anyone's sat in the back.

My phone buzzes again. A second message from Hannah. *'Jock and Touni are mating x'*

Afia peers at my phone and jigs up and down on my knees.

'That's great news,' I tell her. 'You're going to have a playmate.'

Chapter 11

Afia's Home Improvements

Kera had now fully recovered, but as we feared, without Cheery Komi for company she took up her usual spot on the periphery of the group again and plucked all the fur from her shoulders, giving Trim Touni dark looks as she paraded about as head female, strutting her stuff and revelling in her newfound dominant position. Dangerous Sal seemed happy to let Touni get on with it and now Kukena was at the annoying-the-hell-out-of-everyone phase, she mainly ignored her daughter and dedicated her time to stuffing her face. Romina and Afia's relationship was strong, with them both choosing to touch each other through the mesh.

I join Bob in den three after lunch to help finish roping escape routes for Afia, adding baby-gorilla-sized rope ladders all over the house, so if she gets separated from Romina, once in with the rest of the group, there will be no dead ends where she could get cornered.

The gorillas have access to dens one and two on the opposite side of the house, the vast drop slides providing a mesh barrier between us and them.

'I've forgotten how to do a bloody crown knot now,' Bob says. He sits on top of the glass above the public area, a length of rope in front of him.

Afia sits next to him, pats at the glass and the adoring members of the public below.

'Chuck it here,' I say.

Bob snakes the rope over to me. He has separated the end into three strands and stuck a collar of green electrical tape around it, to stop any further unravelling. 'Is it behind or in front?' he asks.

I choose the middle strand and make a loop. 'The left-hand one goes over the top,' I say, 'the right-hand goes behind it, so it's through the loop.'

'Where you off to, Munchkin?' Bob lies back to corral Afia as she trundles off toward the edge of the glass and the long drop to the floor of den three below.

I gently pull the three ends tight, slowly in succession and hold up the crown knot, tightened all the way down to the collar of tape, the three strands dangling and ready to be spliced back into the rope.

'Think we can tick off that she's not bothered by walking on glass.' Bob grabs her ankle and pulls her gently back toward him.

Afia rolls on her back and play-bites his arm.

All three of us have circular bruises up our arms. Not just one or two, we're covered, a jaguar skin of bite marks, all colours of the bruise spectrum; deep purple, yellow and black. We're pushing to get her up to speed and wrestling with her is part of the daily routine, but without thick gorilla skin and fur to soften her play-bites, they bloody hurt.

'Oww,' Bob howls.

'I'll splice the first round in for you,' I say.

Bob rolls with Afia jumping on his head. 'Get off.' He laughs. 'No wonder Hannah's palmed you off on me.'

I pick up my fid, a blunt spike designed to splice rope. Fids are long and conical; the point allows you to spread apart the rope and thread the three divided strands back in on themselves again. One after another, tightening each round as you go. Over and under, over and under to form a back splice, meaning the gorillas can't unravel the rope themselves, pull it apart into tiny strands and accidently hang themselves overnight.

Hannah joins us, another coil of rope over her shoulder and a black bucket of steel rings. 'The knots look good...' She holds down a yawn, her face drawn and pale, having just pulled a double night shift with Afia. 'The hand holds I mean.'

Afia climbs off Bob's head and toddles over to look in Hannah's bucket.

I pass his rope back to him.

'I'm going to start on the ladder,' I say.

The ladder is stainless steel and leads from the floor in den three up on top of the glass hide where we all sit above the visitors.

'I got you this thin rope to use for the ladder,' Hannah says. 'Let's all move down there, give the public what they want.'

We decamp from the glass hide to the floor of den three and Afia potters about between us all, to cries of wonder from the public, pointing fingers, children held aloft, camera flashes and a rank of mobile phones.

The floor in den three has been bedded out with deep straw, so Afia can scale the climbing frame with a soft landing and swing from a set of three ropes we've spliced in as drops. 'Drops' in zookeeper speak refers to ropes that hang vertically, from a long horizontal length. We've made them for Afia to practise on; they dangle in a line, sliced into the rope above, and each one has four evenly spaced knots for Afia to grab and hang from.

'You seen Touni over there?' Bob says. 'Swanning about like queen bee.'

I glance through the windows out into the public area to look across into den two, but it's thronged with members of the public trying to catch a glimpse of Afia.

Bob sits in the straw with her and holds one of the drops steady so Afia can scurry up it. She grips each knot in turn, hands and feet all sure in their grasp, and clambers to the top. She stands with her feet on one of the knots, hangs off the rope with one arm and swings about. Pats her chest with her free hand.

The mobile phones wave outside in the public area.

'We made the right call with Romina,' Hannah says.

I nod. 'Kukena and Touni would be a nightmare if we'd managed to get Afia back with Kera.'

'A right couple of bitches,' Bob mutters. 'Grabbing her every five minutes and his majesty would be beating Kera black and blue.'

'They'd both have Sal on side too,' Hannah says. 'Touni's playing them all just how she wants it.'

Bob snorts. 'Romina and Sal don't give a shit anymore,' he says. 'They don't want to be the centre of attention, barren old crones.'

'You turn into a real arsehole when you're tired,' Hannah snaps.

'Hark who's talking,' he shoots back.

Afia swings from the rope, waves her arm and looks from Bob to Hannah and back again.

'Don't have a row in front of the baby,' I say.

The stainless-steel ladder has four rungs, each spaced at a height that is taller than Afia. She could grab the lowest rung, but to reach the next one up she'd have to balance on her legs, with both arms outstretched and up on tiptoe and there's no way she can do that in a hurry. Particularly if she's being chased by Kukena. I try not to think she could be running for her life. That's a thought we're all trying to avoid, if the introductions go horribly wrong.

Afia lets go of the rope and flops into the straw. She instantly scurries up it again, slapping Bob as she passes. She has her big grin play-face on, all teeth and wild eyes.

Hannah's thin rope is a dream to work with, the strands soft and supple. I splice one end to the lowest rung on the ladder and make sure it's all as taut as I can get it. Kukena is going to be a pain, no doubt about it, but we've provided escape routes everywhere. I obsess over how tight I can splice to the metal ladder, make sure there's no way Afia can get a hand caught between the rungs and rope. Halfway between each rung I add a handhold knot for her to scramble up.

'First mix tomorrow then,' Bob says as Afia dangles from one arm again and kicks her feet at him.

It's the talk we haven't collectively had yet.

'Don't,' Hannah mutters. 'Not if you're going to be an arsehole about it.' She is splicing a ring onto the end of a long strand of thick rope that we can hang from the mesh ceiling high above.

'Romina's ready,' Bob states. 'Does all the behaviours we need her to.'

We all know Afia is ready too.

We've taken her out onto the island, through the external slides, shown her the routes from inside to outside and back again, pushing aside the chewed heavy plastic flaps that insulate the house during winter. Made her watch us grip the electric fence, having turned it off, and terrified her as we shrieked in feigned agony. She happily takes her daytime bottle feeds through the mesh, hangs on one side as we feed her from the other. She is eating solid food and learning to swing from one rope to another, honing her coordination skills and more importantly her behaviours are as they should be for a baby gorilla of her age.

I know we've all done everything we can, but I share Hannah's trepidation. None of us say it out loud, the unspoken horror that we could see her killed right in front of us, hurled against a wall or thrown into the moat to drown.

Hannah digs at the rope with her fid. 'Romina is ready,' she agrees, 'but Afia's still so tiny, that's all.'

'How's the box training going?' I ask.

'Good.' Hannah's deft fingers work at the rope. 'I'll show you it in a bit.'

Bob laughs and tickles Afia's belly as she continues to swing from one arm. 'Which one is it now?' he says. 'Baby number four?'

Hannah nods. 'Kukena pulled baby number three through the mesh yesterday.'

Bob doesn't respond as we all contemplate what could happen if Kukena managed to grab one of Afia's arms through the mesh. She's strong enough to tear it off if she really went for it.

Bob swings the next dangling rope in line toward Afia and she reaches out for it with her toes.

I tighten the knot around the next ladder rung. 'How many cuddly toys have we got left?'

The gift shop donated a selection of stand-in cuddly toys for Romina's training and also gave us the two Clives, the gorilla toys for Afia to play with. Home Clive, tied up in a black binbag to stay clean for his home visits, has been accidently thrown out with the gorilla shit. Zoo Clive is still about, but his inanimate nature has downgraded him from playmate to pillow.

'We've got a red panda toy.' Hannah cuts off the tail ends of each spliced strand with her multitool. The piece of kit all zookeepers get given as a present somewhere along the line, the folding penknife, with pliers and screwdrivers, the tiny saw blade, all the essentials. 'And maybe one more ring-tail.' She has plaited the strands back into the body of the rope four times, the accepted number to stop it unravelling, even if Kera decides to give it a try.

'That red panda looks like roadkill,' Bob says.

We remove the bead eyes from all the cuddly toys, in case Afia pulls them off and swallows them. We get her to play with them, so they pick up her scent for training Romina.

Bob pushes the dangling rope Afia's way again and she manages to grab it this time. She grips the rope with her toes and reaches an arm out for it. 'Do the funky gibbon,' he says.

Afia stretches out her fingers and lets go of the rope with her foot. It swings away again and she waves an arm in frustration.

'Not yet,' Bob says to her, 'wait for it to swing back. Now. Grab it.'

Afia clutches the rope with her toes, makes an instant grab with her hand and clings on as she swings, climbing from one rope to another.

'Swinging winning.' Bob shuffles away from her in the straw, waits at the next dangling drop in line.

Afia slaps her chest in self-congratulation. Mouth wide.

I measure my rope, enough length to surround the top rung and splice it back in. Bite off a strip of green tape to mark it up, to make sure I don't cut the rope too short.

'Shall we clear everything out?' Hannah says, with a glance at her watch. 'We can hang this long one up tomorrow.'

'I'll be five minutes on this,' I say.

Sal and Kukena both knock on the drop slides in unison. They know it's time for their tea, remind us with a series of clangs.

'Cool,' Bob says and leans forward on his hands and knees, offering Afia his back.

She climbs aboard and glances at me over her shoulder as Bob picks up his fid and heads for the door. 'I've got one of the knives,' he says.

'I've got the second one here,' I confirm.

Hannah chucks her fid into the bucket of metal rings and D-shackles. 'I'll sort diets,' she says to me. 'Let me know when you're done and locked up. We'll train Romina when she's finished her tea.'

Romina sits patiently in the end feeding den, where we knocked out Cheery Komi all those months ago. She has finished her tea and grumbles encouragement as Hannah carries a plastic crate and the red panda cuddly toy into the next-door den.

The rest of the group wander the house, going over each other's leftovers. Kera clutches lettuce to her chest as she sidles outside with four carrots held in her teeth.

Kukena is ready and waiting to be a pain, standing up to peer at the vulnerable blind red panda toy.

'No,' Hannah states firmly and makes sure the plastic crate is out of reach. Kukena's arm is still just small enough to wedge through the mesh, not quite up to the elbow, but her long palm and fingers can reach further than you think.

Romina sits on her heels, shut into her feeding den, and pats her round belly with a gnarled hand.

'Nearly ready, Romina,' Hannah says and steps back out into the keeper-corridor, leaving the plastic crate in the den next to her and locking the door.

'We're going to need a buffer den between them for a while, aren't we?' I say. 'Once Afia's in.'

Hannah nods as she picks up the remote control. She sits cross-legged in front of the keeper-door to Romina's feeding den and sets up the training station with a dummy bottle of milk. She wears her training pouch and has a squeezy juice bottle with a long straw, filled with Romina's favourite fruit tea.

'Opening the slide, Romina.' Her training voice is clear and calm.

Romina waits for the slide to fully open and strolls through.

'Good,' Hannah says. 'Get your baby.'

She pulls the plastic crate toward her and sniffs the cuddly toy.

'Get your baby,' Hannah repeats.

Romina lifts the red panda out of the crate and brings it back through the slide into her feeding den. Drags the toy by its red-striped tail.

Hannah frowns. 'She carried the ring-tail more like a baby.'

'Afia will be holding on too,' I add.

Kukena jams her arm through the mesh, as far as she can, fingers outstretched, but despite having been moved, the plastic crate remains out of reach.

'Good.' Hannah injects some enthusiasm into her tone, reaches into the training pouch threaded onto her belt and gives Romina a grape.

'Hold,' Hannah says.

Romina sits in front of her and lifts the red panda up to the mesh with one hand.

'Good.' Hannah posts another grape into her waiting mouth.

Romina grumbles.

'Hold.' Hannah picks up the squeezy bottle.

Romina takes the straw in her lips and continues to keep the cuddly toy at the mesh.

'Hold.' Hannah picks up the stand-in bottle of baby milk and holds it to the red panda's down-turned mouth. She rewards Romina with squeezes of fruit tea, keeping her in place for a full 2 minutes, the time Afia is now taking to neck her milk feed.

'That's amazing,' I say quietly, so Romina doesn't get distracted.

Hannah ends the simulated feed with the red panda when the fruit tea runs out. 'Box,' she says.

Romina puts the cuddly toy down and pats her belly.

'Box,' Hannah repeats.

She drags the red panda back next door and coughs at Kukena, who is still lurking with intent.

'Good,' Hannah says. 'Box.'

Romina pulls the plastic crate toward her and deposits the red panda.

Kukena sees her chance and hooks the crate against the mesh, grabs the red panda by its front leg and yanks it through the mesh.

Romina grabs its tail and coughs at Kukena. The toy rips and jerks open, torn apart like paper between their vicelike hands.

'Kukena, NO,' Hannah barks.

She bounds off, the blind red panda head and arms twirling above her head.

Romina sniffs the bisected toy and drops it in the box, picks a piece of stuffing off the floor and comes back through into her feeding den.

'Shutting the slide, Romina.'

The mesh door grinds closed.

Romina posts the clump of red panda stuffing through the mesh.

There is no doubt that Romina is as ready as she'll ever be. We've trained her to give Afia up if she suffers some horrible injuries that require vet attention. There is no need to put the initial mix off any longer. Romina and Afia's physical relationship has developed as far as it can with mesh in between them.

It is time to mix them. Reverse the physical dynamic. Once Afia is mixed in, it will be us that only interact with her through the mesh. We'll be bottle feeding her for years, as gorilla babies don't give up their mother's milk until aged three or four, so we will all still be sharing a close bond with Afia, but if all goes to plan our physical connection with her is coming to an end. There will be a barrier between us. Forever.

The prospect hits me like a Sal punch.

Captivity.

The question of what the hell I am doing to these animals, amplified by what we are about to do with Afia, rolls over me for the first time in however many years I've been a zookeeper.

We are about to condemn Afia to a life of captivity is how it feels. I have actively bonded with her, provided her with a family to grow up in. She's become a member of my own family at home and now, we have to work through letting her go and untangle those emotional connections.

The thought that I am denying Afia her freedom is irrational. She can't stay living with us of course. We can't hit the road and live free. She is a gorilla, and I am a human. Afia needs to be back with her own kind.

Wild western lowland gorillas are the most numerous of the four different sub species. Estimates put their number at around 316,000. The mountain gorilla in comparison is down to about a thousand individuals, although it should be pointed out that this number is an increase from the 600 or so that Diane Fossey first met in 1963 and they have increased in number through her pioneering and direct hands-on conservation methods.

Do I want to see Afia introduced to the wild instead of live with the family group at the zoo?

The reason zoos focus on the western lowland gorilla is because the wild numbers provide the best chance of preserving this sub-species from extinction. That's not to say zoos have abandoned the other species of gorilla, all much lower in numbers, but rather they do not keep them in captivity. These sub-species benefit from numerous conservation projects, led and supported by zoos around the world in an attempt to preserve the gorillas' natural habitat, but collectively zoos have chosen to focus all captive gorilla resources on preserving the western lowland gorilla, which acts in part as a population back-up, to hold off the threat of extinction for as long as possible.

Don't misunderstand me here. Animals are not taken from the wild to be displayed in zoos anymore. Those days have long gone. Animals in zoos are almost always born in captivity and if an animal were to be taken from the wild, it would be to save that species from immediate risk of being wiped out completely.

Attempts have been made to reintroduce captive gorillas into the wild, but like any reintroduction program they are beset with a range of complications, one of which is available gorilla habitat.

We've all seen the TV documentaries and experienced the waves of depression at the state of the planet, watching the destruction our own species has wrought upon the world, the environment that we share with the gorillas and all the other species of primate. The rampant destruction of our own environment, be it through carbon emissions and pollution or all-out warfare, is not a trait shared by our primate cousins.

It's rare that any primate species goes to war with its own kind. There appears to be a direct correlation with group size where it does happen. The bigger the groups, the more prone they are to fighting with rivals, generally over limited resources. Food, meaning habitat.

Groups of baboons, macaques in India, all fall out and attack each other, however the comparison with humans is marked in that we, as a species, throughout history have always chosen to be at war. The closest example seems to be in chimpanzees, closest to humans in DNA, who will go off and kill members of different groups. Seemingly as a way of reinforcing social bonds between themselves, they'll get together, follow the head chimp off into a rival group's territory and brutally murder one of them. The point being, the wild for all species of primate is a shrinking, dangerous environment beset with human warfare across poverty-ridden, starving communities. Snares and the bushmeat trade. Disease and the illegal pet trade, the logging industry, mining and the relatively recent eruption of palm oil plantations. The list goes on and so for me, the answer to the question would I rather see Afia eventually released into the wild, is no.

I agree with the stated aims of zoos. We should find out all we can about the species we hold, the natural world and how everything connects. Fight to understand and protect the dwindling habitat that after all produces the very oxygen we all need to survive. Everyone on the planet.

There you have it. My method of morally justifying why I am a zookeeper.

Afia would be kept in captivity for the rest of her life, but she would experience natural gorilla behaviour. Getting her into the group now, at such a young age, would mean she'd learn to understand the society she was part of, so we wouldn't be creating another Kera, ostracized and alone.

We'll finish roping the house tomorrow morning, during the cleaning routine and then we'll have no reason not to try a mix. Afia is seven months old and when I take her home tonight, she will climb out of the car seat and get on my lap. Cause two dogs in the four-by-four in front to go berserk as she peers over the top of the steering wheel at them.

Chapter 12

Last Night at Home with Afia

Parents will recognize the feeling of fear when the kids first go off and do something dangerous. Wobble off on their pushbikes, swarm into the sea with bodyboards. Call out from the top of spindly trees, turn up delighted that someone has given them a skateboard with wonky old wheels. The terror if you're unlucky enough to have one stay in hospital. As the kids get older, the cold ache left behind as they stumble out into the world as young adults, when your care is no longer required, a change within the family group structure for all concerned.

Knowing tonight is the last time Afia will be with us means none of the usual playing has the same joy, tinged as it is by the fact it will not happen again. This is the last time Afia will try and pull the wifi box out of the wall and set up the coffee table and cushions just how she likes them. She has given up pulling the cushions up and down the couch and has taken to chucking them all on the floor instead, making a pile around the coffee table, which she uses as both her route up onto the couch and a launch pad. The first hour at home is spent endlessly standing up on the coffee table and patting her chest. Over balancing and falling onto the cushions, rolling to her feet and back up onto the coffee table again. This progresses into jumping onto me off the coffee table and wrestling on the floor. I lie on my back, fend off her play-bites, to get my mouth into the crease of her neck, nibble and get her hysterically chuckling. She does that over and over, interspersed with running about the lounge throwing plastic bottles around and making surreptitious grabs for the wifi router.

Sparky tries to hide her sadness that this is Afia's final visit home behind a wall of practicality. 'We always knew this day would arrive.' She winces at her own cliche. 'I can see now what sort of dad you would've been with the kids as babies,' she says as I roll around on the floor with Afia.

Afia is too strong and bitey to play with Sparky now. But when she wants quiet time, she chooses Sparky over me. Whereas I am always awoken by repeated slaps on my forehead, Sparky receives gentle strokes to her sleeping face. It's worse for Sparky in a way. I will be seeing Afia every day at work still, but for her this is it.

'Do you think Romina will take her?' she says.

'It might be a week or two yet,' I reply. 'There's a zoo flat over the road available now, they've been doing it up, so we'll have to take her there overnight, now she can climb out of the car seat.'

'I thought once you put her in with Romina that was it?' Sparky says.

'Tomorrow we're going to let Afia go in with her. We'll get Romina upstairs in her feeding den and set up with Afia next door.' I can see I've lost Sparky with all the talk of dens.

Afia trundles over to her toy bag and tugs out the tubular wire lampshade. She bashes it with an empty plastic bottle.

'We'll open the door between the two dens to a width that Afia can get through, but Romina can't.'

'Romina could get her arm through that gap, though?'

'Yeah. We've drawn safety lines on the floor, in case she tries to grab us. But she won't. She's lovely.'

'Be careful,' she says.

'It means Afia can come and go between us. We want to see Romina pick her up ideally.'

The front door sounds and Afia responds as normal. She stops her bashing of the lampshade, plastic bottle paused in mid-air and looks up at the lounge door expectantly.

Fizz appears. 'Hey everyone.'

'It's Afia's last night with us,' Sparky says and squeezes my hand.

Fizz frowns. 'Shit, Al are you okay?'

Afia jumps up and down on the lamp shade. It crackles and buckles beneath her strong legs. She waves the plastic bottle.

Fizz crouches. 'I can't believe it's your last night.'

Afia sees Fizz on her level and bounds across the lounge, throws the bottle over her shoulder. Arms lifted, play grin across her face, mouth full of teeth.

Fizz jerks backward, face frozen.

Afia reads the look of sudden fear. Glances over her shoulder to see the threat, as she can't conceive Fizz is scared of her, and panics that we're all in mortal danger. Runs at Fizz to be swept up and saved. Screams in fear.

Fizz sees Afia on the attack, screaming in rage, a shrieking mass of muscle and bitey teeth and bolts out of the lounge.

Afia follows and they both break into the same ear-piercing shriek. I leap up after them. Fizz on the dining room table, feet running on the spot, Afia charging round and round in terror.

Afia leaps up into my arms. Her heart pounds, arms clamped around my chest.

Sparky bounds into the kitchen after us, driven by the shrieks from Fizz. 'What are you doing?' she says to Afia, her tone stern. 'Are you okay, darling?'

Fizz dismounts from the table.

Afia clings to me and peers over her shoulder, shoves a thumb in her mouth.

'She wasn't trying to attack you, Fizz,' I say, 'she thought something was coming to get you both, she was running off with you.'

Fizz breathes hard and makes a massive effort to calm down. 'Her teeth were all bared.'

Sparky glares at Afia.

'She wasn't being aggressive,' I reassure them.

Later we share our last meal together. Afia is subdued, sensing the flat mood around the table. She chews her way through long leaves of bottle green chard and licks most of the tomato seeds off her fingers, before wiping them on the edge of the table.

'It feels like only yesterday when she first came back,' Fizz says, doing a good impression of being recovered from the chase, after I'd explained again why Afia had behaved that way. Recovered enough to sit at the kitchen table with her at least. 'How do you think it'll go?'

Afia gives me one of her intent looks.

'I think it'll go okay.' Afia reaches up to touch my lips. 'Romina is such a calm, lovely animal.'

'Time to grow up and be a gorilla,' Sparky says.

Afia sleeps through the night, on her back between us, a hand on each of our arms. I slide out of bed, sneak down the stairs to prep. I'm not sure how I'm going to get her into the car seat unless she's asleep.

I decide to sack off breakfast and settle for a coffee at work, creep downstairs and open the front door as quietly as I can. I hear the first thump from upstairs as I step outside to load the car. She's awake and has begun dropping the books, piled up on my bedside table, one by one onto the floor.

'Shit,' I murmur.

I need to grab the car seat, didn't bother bringing it in last night as Afia was already onboard. Zipped up inside my hoody in a mass of bulging arms and legs, head trying to push out and have a look.

I'm parked outside Lorna's house. Her front door is already open and she comes pattering across the road behind me, from Janet's place, both of them up and about and it's not quite six in the morning yet.

'How's your little daughter getting on?' Lorna says with a laugh and playful nudge.

'It's her last day coming home,' I say and surprise myself at the lump in my throat.

Lorna rubs my arm. 'Bless her, going back in with her mum.'

'A surrogate,' I say. 'Lovely gorilla.'

'You'll miss her I expect.' She rubs my arm again.

'I'd better go grab her,' I say. 'Might be a while.'

'Oh,' Lorna whoops. 'Is she tearing about?'

I carry the car seat back indoors and trudge upstairs to the bedroom.

She's lying next to Sparky; they have their faces close together. Afia gently strokes her cheek, peers at her eyebrows. The books are scattered on the floor, but Sparky has got her to calm down again.

Fizz comes to say goodbye. I can see the chase from last night is still fresh. 'Bye-bye, Afia.' Fizz rubs my back. 'Good luck today, Al.'

The car seat provokes full shrieking melt downs, a wriggling free of straps and desperate attempts to get back into bed with Sparky.

We're all crying.

I give up on the car seat altogether. I drive carefully, with her zipped up inside my hoody where at least she feels snug and safe, which makes her slightly less wriggly. The roads are quiet this time in the morning, and I breathe a sigh of relief as I park up behind the usual gate.

We set up in the upstairs feeding dens. I sit cross-legged in the den next to Romina with Afia on my lap, making sure I'm behind the line we've drawn on the floor with a Sharpie. The line is measured at a distance that should be outside of Romina's arm. Even if she lies on the floor and jams it through to the shoulder, fully extended, I should be out of grab range.

Afia pats my wellies.

Bob feeds Romina through the mesh. 'Are we all ready?'

'Yep.' I tickle Afia's belly.

'Me too.' Hannah picks up the remote and I watch her eyes flick across the buttons as she checks repeatedly what she's about to do.

'Good luck everyone,' Bob says. 'Mix time.'

The rest of the group are shut into dens one and two. Romina is on her own and the adjoining den on the opposite side, where Kukena likes to lurk in the hope of grabbing a cuddly toy, is also shut down. None of the other gorillas can get to any mesh that Afia could shortly be pushing her arms through.

There is a second Sharpie line drawn above the slide that links Romina's den with the one me and Afia are sitting in. It's measured to accommodate Afia getting through, at high speed if she needs to.

'Opening the slide,' Hannah says.

I get the flush of adrenaline as the slide grinds open, to reveal a gap between me and Romina.

She gives me a surprised look. No barrier of mesh between us. Registers it's unusual, but Bob feeds her another piece of chopped apple and she seems to accept it as fine. Doesn't go mad and throw her arm through the gap in a frenzied grab for me or Afia. I force myself to be calm, relaxed, double-check I'm behind the Sharpie safety line on the floor.

Afia scrambles off my lap and toddles through to see Romina almost straight away. Makes her jump when she barges into her.

Romina makes a grab for her with a long hand and groping fingers.

Afia ducks away.

Romina holds her hand to her nose and smells her fingertips.

Bob keeps the treats coming. 'Good,' he says to her, his voice level and calm.

Afia scurries around the feeding den and then wedges in next to Romina to see what Bob is dishing out. Clambers across Romina's belly, to pat at him through the mesh.

Romina reaches for her again with both hands, but Afia pushes her way loose and clambers all the way around her to trundle back through the gap in the slide to see me.

I don't give her eye contact, remain still and as unobtrusive as possible.

Romina grunts and Afia heads back through and rolls over on the floor behind her back. Picks herself up and pads about the den again. Her pace is off, the only sign she's on edge. Quicker than usual, which in turn makes her jerky and uncoordinated.

We collectively hold our breath. Bob has his hand in the treat pouch, ready to distract Romina, but doesn't offer her anything.

Romina swivels about to watch her.

Hannah has the remote control and I'm in place to scoop Afia up if she flees.

Afia ducks away each time Romina throws out an enquiring hand.

'Is she trying to get out?' Bob says.

'You can see her pace is off.' I peer through the slide.

Romina grumbles at us all again, but whether it's focused on Bob and his treat pouch or Afia, it's hard to say.

We end the first mix when Afia comes back into the den with me. We leave everything on a positive and I bring her out into the keeper-corridor.

Hannah lets Romina back out to be with the rest of the group. We're running a balancing act. Romina is great at coming in when you need her to and enjoys training, but like any primate she doesn't want to be shut away on her own for too long. The plan is to build up the length of time these 'soft' mixes last and see how Romina and Afia's bond develops.

Afia is exhausted after the mix and once she's had her bottle, she falls into a deep sleep. The group are all asleep too, digesting breakfast.

'Let's get coffee and figure out what we're bringing to this meeting later,' Bob suggests. 'With the Higher-ups.'

Hannah nods. 'We need to be able to lay out some reality, explain what's going on.'

'And everyone's asleep in there,' Bob says. 'Afia's not missing out on anything, is she?'

I carry the sleeping Afia out into the messroom.

'Heard about the car seat,' Hannah says to me.

I sit with my legs up on another chair, Afia lying on my thighs. 'Forced herself out of the straps,' I tell her, 'climbed straight out.'

'Had a chat with the Higher-ups about that this morning,' Bob calls from around the corner by the coffee machine. He's microwaving coffee for us, nuking yesterday's cold sludge.

'We've got sign-off to use the basement flat over the road from now on.'

Bob hands Hannah a coffee. 'They're waiting to redecorate it, that's why it's empty, so don't matter if she trashes the place.'

Hannah wrinkles her nose. 'You make it sound so charming.'

'Ain't going to be long before we leave her with Romina overnight, is it?' Bob says and passes me Twilight Scott's sloth mug, the back legs and claws making up the handle.

'Or it could be weeks,' Hannah responds, and her tone has a touch of hope about it.

'We need to figure out how to run the mix,' Bob mutters, even though we've been talking about nothing else. 'Going to have to be fluid about it, could be Romina takes her and she's okay. That's when we step out. Leave them to it. Stop.' He sits down and scans the table for scrap paper and a pen to write down our thoughts for the meeting. Lifts the section diaries and dislodges a pile of old wildlife magazines.

'We need Romina to pick her up,' Hannah says. 'But Afia has to be at a point where she's happy for that to happen.'

Bob unearths a blank temperature sheet to write on.

Afia shudders. Dreams. Her eyelids flutter and she sighs. I take the pen out of the hand-rearing bag and pass it to him.

'Nice one.' He flips the sheet of paper over.

'She's not scared of her, though,' I say. 'That means she must trust her, right? Romina is familiar to her.'

'Didn't like Romina pawing at her, though.' Bob picks up one of the diaries to lean on.

'No. But she didn't look panicked, just shoved her way loose.'

'Do you reckon Romina would pick her up on the *Baby* command?' Bob asks us both.

'Not yet,' Hannah says. 'We're going to have to work that in later.' She looks up at the monitor on the wall. 'I wonder if this time of day works best to mix them? After a feed, when they're all sleepy. If we could get Afia to sleep on Romina, we'd probably be as good as there.'

Bob scribbles away in his illegible handwriting. 'Al, I think we do like you were doing, when you was in the den. Being as invisible as possible, not providing a distraction.'

I nod. 'No eye contact.'

Hannah blows on her coffee. 'We can start edging further away.'

'Stop being someone to hang out with,' I say, much as the thought drives a sudden cold tip into my relationship with Afia. 'We ignore her when she comes in to see us. See if Romina will play with her. If that happens, we could be as good as there too.'

Bob nods and scrawls more loops across the page. A few arrows and something underlined.

'Prepare for a slow introduction,' Hannah says, 'but explain we're going to be led by Romina and their relationship.'

It's the physical contact between them that we're after. Hugs. The people we hug in our lives are generally our closest friends and family. The instinctive joy of hugging someone you love, the protective shared strength is a common behaviour between humans and gorillas alike.

It takes a week to reach the point of leaving Afia on her own with Romina. The soft mixes work okay, Afia's confidence rises, but Romina clearly believes she is our offspring. She respectfully doesn't try and pick her up, pin her down and sniff her, or respond when Afia instigates play. She lets us feed Afia, who toddles to the mesh on her own accord. Romina accepts the fruit tea and never tries to push Afia off her milk.

We encourage them as much as possible, but it is clear that our presence during the mixes is not contributing toward the desired results. It is time for the next step; to leave them together alone.

My final night with Afia is spent in the bleak basement flat over the road from the zoo. It is cold and dark. After some half-hearted play and pulling the curtains down, she has her bottle and becomes heavy with sleep. She sleeps through the night with her head on my chest, snuggled up against the cold.

Bob takes the lead on leaving Afia alone.

None of us want to do it. Abandon her. An act that feels weighted with betrayal. Bob spares us and volunteers.

We gather in the messroom, watching him up on the monitor. We'd been moving closer to the keeper-door during the soft mix. Progressed to sitting outside, even closed the door part way, but we'd never left Afia alone.

Bob waits for Afia to amble through to see Romina and steps out of the den. He closes the door. Afia clocks him and flees through the narrow gap between the dens.

Hannah operates the remote from out here in the messroom. She opens the dividing slide fully, so Romina has access to both dens.

Afia shrieks and tries to force herself under the gap beneath the keeper-door to get to Bob, desperate and terrified.

Romina kicks off. Jumps and shrieks, casting about for the source of panic.

Bob crashes out into the messroom, stumbles over a chair, eyes filled with tears, fixed on the monitor.

We gape, faces frozen in shared panic as Romina screams and coughs.

Afia, frantic, runs from one den to the other. We see Romina, up on the screen, appear to lunge for Afia and snarl, the moment we've dreaded poised to engulf us. A sharp slice of terror, a blast of cold joylessness that will haunt us for the rest of our lives, that this intense, surreal and magical experience was about to end in the worst, most violent way, right in front of us.

We brace ourselves to charge back in, but Romina's shrieks and screams stop.

Afia has shat everywhere. She stumbles about, anxious, her movements jerky, and Romina reaches out for her.

We hold our breath as she lifts Afia onto her belly and pats her back with a heavy gorilla hand. Afia's head turns from side to side, but Romina continues to rub her back and we see her begin to relax. Watch her hands clutch the fur on Romina's chest and arms.

'We've done it,' Bob whispers.

Their hug changes everything, and, in that instant, our physical contact with Afia is over. We still have to reintroduce her and Romina back to the group, but even if that goes wrong, or takes months on end, Afia will never now come out of the gorilla house overnight. A new phase of our relationship begins, the boundaries redefined. We are no longer the parents.

That is now Romina.

Afia and Romina begin to play together. Romina encourages Afia to jump on her head and they both erupt into synchronized hysterical chuckling fits. At night they share a nest and Afia begins to ride on Romina's back. Hands on her shoulders, fingers and toes clenched in Romina's fur. They are at the stage where we can try introductions to other members of the group.

Dangerous Sal is first. Barrel shaped, with her scarred pointy head. Battle worn and the most experienced female where babies are concerned. She strides straight over on her thin stubby legs and drops to her elbows to take a long stare at Afia, clamped to Romina's wrist.

Afia's eyes are huge.

None of us, Romina included, like the way Sal is staring.

She seems intrigued that somehow Romina has managed to get hold of our baby and appears to be making up her mind as to whether she wants Afia for herself. She knows the potential rewards we might dish out if she were in charge would be astronomical and perhaps, at last, this would lead to her long-awaited breakthrough in communication with us, that she's been tirelessly working toward all these years. The simplest command, the equivalent of: 'You there. Baldie. Give.' If she had Afia, would we do literally anything she required to get her back?

I'm on my toes, ready to do what? I can see the thought flashing across both Bob's and Hannah's minds. We are powerless to intervene. My heart bangs in my ears.

Sal reaches out to take Afia.

'NO, SAL,' the three of us shout in unison.

Romina goes berserk.

Romina rarely loses her temper, so an attack from her always has the added element of surprise. She snarls and coughs, lunges at Sal and bites her shoulder. Rains blows down on her head and back as she flees, shrieking in rage.

Jock goes ballistic next door. Unable to get in there and admonish Sal for upsetting Romina – his favourite female – he chases Kera outside instead.

Once Romina loses her temper, she always goes for a whole additional round of attack, more than is strictly necessary or deserved, but her technique leaves no one in any doubt about the consequences.

She chases Sal into a corner, Afia clinging to her wrist, and delivers another bite to her back and round of thudding blows.

Sal cowers and screams.

The rest of the group get the message.

Trim Touni, having witnessed Romina's attack on Sal, didn't try to interfere. Her own pregnant belly was beginning to show, not that it stopped her trying to get Kera bashed up, but she adopted a respectful and wary approach to Romina's new baby.

Kukena remained a pain for the first few months, trying to grab Afia if Romina took her eye off her, until the day the slides jammed in the isolation den.

It was during the feeding routine. Kukena pushed her way into the isolation den with Romina and Afia, in an attempted snatch and flee.

Afia trundled out of her reach as the slides closed shut and immediately jammed.

The slides for the isolation den are on a different circuit and they froze, meaning we had to run upstairs to reset the system.

Romina took the opportunity to beat the crap out of Kukena, twice.

By the time we managed to let Kukena out she was a quivering wreck and didn't try to make off with Afia again for at least two weeks.

Kera showed occasional interest in Afia and would reach out to touch and sniff her daughter if Romina would let her, but they had no real physical interactions.

Jock didn't seem to notice Afia existed. He ignores his offspring for the first year or so, not interested until they reach a size where he can play with them. After Afia had been in with the group for six months and was just over a year old, she first approached Jock. Romina sat close by, tense across her shoulders, ready to intervene and haul Afia out of harm's way, but Jock lay on his front and Afia grabbed his lower lip in both hands.

This highly strung, enormous silverback, his huge head the size of Afia's entire body, grumbled encouragement and let her yank his lip this way and that.

We'd done it. Afia was in the group and becoming a proper baby gorilla.

Chapter 13

Death and Misery

Trim Touni made first-time parenting look easy. If she were human and visited the zoo with the rest of the tired mums who gather for support on weekday mornings with their coffee, Touni would have been the energized, breezy one, clad in sportswear, having run to the zoo with a fancy stroller, the baby fast asleep. Trim, showing off how in shape she was again already, much to the envy and loathing of the exhausted coffee drinkers, with their talk of sleeping regimes and potty training.

Touni's baby was another female, named Ayana. She fed right from the get-go and rode around on Touni's wrist in a proper gorilla fashion. A baby gorilla wraps arms and legs around their mum's wrist in the early stage of their development, as they are not yet big enough to ride on their backs. They're tiny, and as the mums knuckle-walk around, they support the infants' bums in their backward-facing palm. Create a little shelf for them to sit on.

As a zookeeper, if you see a baby primate dangling a limb, you know it's weak, not getting the nutrition it needs and is heading for trouble. Ayana's grip is strong, a sure sign that she is feeding well, and she hugs her mum's wrist even when upside down as Touni clambers about. Up and down the climbing frames, crossing the enclosure without going to the floor, swaying across the web of bouncing ropes, Ayana clings on throughout.

As Ayana gets older, Touni casually swings her up onto her head, like she is wearing a hat, not caring which way round her infant faces. Usually, a gorilla infant will face forward, but Touni leaves that for Ayana to figure out and begins to adopt a more hands-off approach, which causes fits of pitiful crying from her daughter. She is wary of Kukena, however, who is at the running-off-with-shrieking-kids stage of her life. Touni doesn't want to risk offending Sal and bides her time until she eventually corners Kukena outside. Before Kukena realizes everyone else is indoors and that all her back-up is out of sight, Touni leaps on her and gives her a pasting.

Jock comes storming outside at the sounds of Kukena's wails, but Touni is quick and efficient, Ayana slung under her belly throughout the assault. She scampers away behind the ridge as Jock barges out through the slide and circles around to reappear behind the shelter. Her movements are slow, as if she had been innocently foraging the whole time.

Jock glowers up at the trembling Kukena, perched high up on the climbing frame, and crashes his way back inside to blame Kera.

Afia has been in with the group for about a year and is beginning to play with Ayana. Both Touni and Romina are relaxed enough to let them get on with it and I see them rolling about and chasing each other through the slides from one den to the next each morning, as I leave the morning meetings in Gorillas and head up to Twilight World.

Now Afia is in the group, I've returned to the small mammal team and have been promoted to maternity cover as team leader. It's more paperwork than animal husbandry and there is a great team, including Twilight Scott. We refit enclosures, plan animal moves. I keep my hand in at Gorillas and drop in to give Afia her midday milk feed through the mesh.

On my day off I get a call from Bob, summoning me to the pub that night as they came into work this morning and found Sal had died overnight.

I stand at the bar and wait for them both to arrive at the pub. It's dark outside already. Late in the year. Past bonfire night, which happily didn't appear to freak any of the animals out this year.

Bob leads Hannah in out of the cold and I can see by their faces they've been rowing on the way to the pub. Bob will be hiding his grief behind the usual wall of flippant digs at Hannah.

'What are you both having?' I ask.

'Cider,' Hannah says. 'Please,' she adds. 'How was your day off?'

I can see she could burst into tears at any moment.

'Fine. How are you two? Bob, what are you drinking?'

'I'll have a cider too, mate,' he says and picks up the bar menu, turns it over and casts his eyes around for the specials board. 'Don't know why I'm looking. Another week 'til payday.'

Zookeeping is not a well-paid profession. Eating out is rare, savoured, and even trips to the pub are pretty infrequent.

We take our pints to the table in the corner, next to the wood-burning stove. Bob brings the menu with him, and we bundle around the table. 'Bloody starving,' he mutters.

'Did you see what happened?' I ask them.

'I'll show you.' Hannah pulls her phone from her jacket pocket. 'It happened around five-thirty last night, just after we'd all left.' She's wearing a new jacket with a fur-lined hood. It brushes my cheek as I lean in to see the screen and the sensation reminds me of Afia.

Hannah props her phone against the candlestick wine bottle and plays a video taken from the rewound camera feed. Sal sits on the hide in den two, in a big wood-wool nest. Kukena is nearby and Sal pulls an aubergine out of her hand, takes a bite and shudders. Slumps over in her nest. A matter of seconds. Robbing her daughter one moment, gone the next. Just like that.

'Heart attack.' Bob scans the menu.

'You were balling your eyes out with everyone else this morning,' Hannah says.

'Not a bad way to go, is it?' He rolls his eyes for my benefit, and I know he's upset of course, but he's right. 'That's all I'm saying,' he adds.

We'd worked hard to get the weight off Sal for years, a constant struggle as she outwitted us and stole other people's food. She was 42 years old, a good age for a gorilla. Seeing the footage rams it home. An animal that had been at the zoo every day, for the entire time any of us had worked there and she'd dropped dead. No lingering illness or veterinary procedures. Just the end of the road.

'Can't believe how quick it was,' I say.

'Would you have the steak and ale pie, or lamb shank?' Bob asks us. 'No wait, pie, lamb shank, or fish and chips?'

It's Hannah's turn to roll her eyes. 'Have a look at the rest of the footage tomorrow,' she says to me. 'Kukena tries to wake her up, shakes her and then eventually they all come over and sit around her in some sort of weird mourning circle.' She gulps and pauses to sip her pint. 'I know she reached a good age, but I'll miss seeing her.'

'Cornish scallops,' Bob reads from the menu, 'with chorizo.'

'The footage was so interesting,' she adds.

Bob looks up from the menu. 'They all gathered round to drink the gunk and gloop that came out of her,' he says. 'Not sure I could face the risotto. Saw one go past just now and it looked well sloppy.'

'You remember how Jock went in and out of the house when we exported Moki to Belfast?' Hannah says. 'He kept going outside and calling for her. None of that with Sal. They know she's gone. Understand she's not coming back.'

'He's miserable as all hell, though,' Bob adds. 'Poor sod, sitting about crying.'

'It was fascinating,' Hannah says.

'I wonder what they think about death?' I sip my pint. 'What they understand about it.'

'Kera didn't give a shit.' Bob flips the menu over. 'Pissed about in den three when they was all sat round her.'

'The group weren't bothered when we took her out either,' Hannah says. 'All weirdly accepted it.'

'Don't stick your head in the freezer up at stores for a while,' Bob warns. 'She'll get her post-mortem off site next week.'

Hannah glares at him, visibly upset again.

'They must have a concept of death,' I suggest, offering it out to avoid a potential outburst from either of them. Hannah looks like she's on the verge of stabbing Bob in the head with the conical saltshaker. 'At what point do they realize she's not going to wake up again?'

'When all them bodily fluids come leaking out of her,' Bob says.

Hannah's nostrils flare.

'What's interesting,' I continue, 'is that Jock is sad, visibly miserable. That's got to mean he's remembering her.'

Bob snorts. 'The good times,' he says. 'Watching her eat gurt ice cream cones of her own shit.'

'Seriously.' Hannah lays both hands on the table.

'Never mind that,' Bob says. 'What's your perfect roast dinner?'

Hannah tuts. 'Really? Again?'

'Come on, Treacle,' he says. 'What Sal would've wanted, talking about food.'

'Doesn't this just make you hungrier?' Hannah asks, resigned, but relieved he's going to lay off Sal for a bit.

'Al? What meat would you have?'

'Lamb.' I take up the baton. Bob thinks talking about food is almost as good as eating it. 'Leg, studded with slithers of garlic all over.'

'Talk to me,' Bob says, eyes alight.

'Paprika and garlic powder rub,' I say. 'Blast it on high for a bit, get a crust going and chop up some veg for it to sit on when you turn it down low.'

'Like what?' Bob sits forward in his chair and finishes the rest of his pint.

'Carrots, parsnips, onions, halved. Tons more garlic. Bay leaves.'

Bob nods. 'Potatoes? When you going to chuck them in?'

'Later. Don't want them disintegrating.'

'See,' he says to Hannah. 'Got to have proper roast potatoes.'

'Bottle or two of white wine in there,' I say.

Bob groans.

'Some stock. All covered over with foil. Turn it down as low as you can get the oven and leave it for five hours. Turn it over a few times. Carrots and the rest of the veg get cooked in the lamb juices and wine.'

Bob grips the edge of the table. 'Keep going.' He gasps and rolls his eyes to the back of his head.

'You're so repugnant,' Hannah says. 'No wonder you haven't got a girlfriend.'

'Drinks?' he says.

I drain my pint. 'Sure.'

'No more food talk, though.' Hannah shoves the menu into his hand as he goes back to the bar. 'Come and see Jock tomorrow,' Hannah says to me. 'He'll want to see you.'

'How's Afia doing?'

'Great. Her and Romina were sat in the shelter at the weekend, side by side. Eating away happily. No trying to prize food out of her mouth...' Hannah stops herself saying something along the lines of *'like Sal would have done'*. 'She's touched Kera a couple of times,' she says instead.

I pick up my beer, forgetting I've finished it. 'Is there any maternal bond there do you think? Between them?'

'Kera's interested.' Hannah frowns. 'But not really, no.'

'How about Ayana? Is Touni letting her feed?'

Hannah laughs. 'We've got Ayana's den set up on a break, so Touni can't get in,' she says.

'Break' is zoo-speak for a slide closed partway, leaving a specific gap to stop older, bigger animals getting in.

'We stick Ayana's diet in there,' Hannah continues. 'Like we do with Romina and Afia. Afia goes in, sits there happily munching away. Romina leaves her to it, right?'

I nod.

'If Ayana goes in, Touni waits for her to pick something up and panics her. Ayana runs out to get to her and Touni just robs whatever she's holding.'

I laugh.

'Ayana's terrified to go into her own feeding den. And if she tries to run out empty handed Touni blocks her way, shoves her back in again.'

'Are you hand-feeding Ayana, then?'

Hannah nods. 'Touni had her by the ankle the other day, trying to fish a lettuce with her own daughter.'

'Three pints of Rattler,' Bob announces and places a triangle of pints on our table. 'Bought dinner as well.' He steps back to the bar and brings us a packet of crisps to share.

I visit Jock the next day and nip in for an hour after lunch. He sits on his training platform in front of the den two keeper-window. I head down the keeper-corridor, instinctively look for Sal in her usual spot at the door to the isolation den.

I grumble as I walk down to Jock, to let them all know I'm here.

Jock leans his huge head against the mesh. Feet curled under him, arms across his chest. His eyes don't hold the usual hopeful look that training treats might materialize. He doesn't move at all.

'Hello, Jock,' I say, tone low and quiet.

He holds the back of his huge wrist against the mesh.

I lightly touch the fur on his arm. 'Life's shit sometimes,' I say.

He sits up and looks about the enclosure, but his gaze lacks focus. His long lips push out and he lets out a low hoot. Quiet. Not a call for the rest of the group. The repeated 'who, who, who', it's a subdued cry I've never heard him make before.

'You're okay, mate,' I say.

He slumps back against the keeper-window again, arms around his knees, shoulder and head leant against the mesh.

I sit with him. Chat.

Eventually he gives my arm a poke with his thick fingers and climbs off the training platform. Wanders across the floor and lies down in front of the huge metal ladder.

I move back down the corridor to see if I can catch a glimpse of Afia at den one window. She is up on top of the glass, Romina lying on her back nearby. Afia holds a long rope in one hand and runs a circle to swing out above the floor of den one in a wide loop and land on Romina. I hear them both laughing.

Jock blocks the slide from next door and grunts when he sees me at den one window.

Romina pulls Afia close up above and keeps an eye on Jock as he hurries across to the nesting basket. He pulls his huge body up in front of the keeper-window and his lips wobble as he lets out his low hoot again.

I grumble and he adopts the exact same pose from a few moments ago in den two. Sighs and holds his wrist to the mesh.

I lean close and we sit together until he shoves my arm with his giant fingers again and clambers out of the nesting basket.

'How are you getting on?' Hannah calls down from the upstairs keeper-corridor.

'He's sad.'

'Do you want to feed Afia her bottle?'

'Hell yes.' I miss Afia of course. She's been in with the group for well over a year now and we've been limiting our interactions, trying not to undermine the relationship she has with Romina, but we still need to bottle feed Afia three times per day. Precious moments when you can look into her eyes and have her pat your hands.

I head upstairs.

'Hello, Jocky?' Hannah says as I step into the keeper-corridor. 'You looking for Al? He's here, look.'

Jock climbs up the steep gorilla stairs and strides into one of the feeding dens. Leans his bulk against the keeper-door.

I crouch next to him. His lips wobble, but he doesn't hoot. His eyes fixed in front of him, not taking anything in.

'Made up with Bob yet?' I ask Hannah.

She looks sheepish. 'He had that coming.'

I nod. 'Yeah, he was on one last night. Had you been anywhere else on the way down?'

'He drank his usual cans on the way.'

'It's just his way of coping.'

Jock offers up the back of his wrist.

'Good lad,' I tell him.

'I understand that, Al,' Hannah says. 'But it doesn't always go hand-in-hand with my way of coping.'

'Did he say sorry?'

'Yeah. After I'd stopped yelling at him. He can be such a prick and he knows how much I loved Sal.'

'You loved her, didn't you Jock?' I say.

'Bob said we should all go out for a real meal after payday. Actual food, rather than imagined.'

'Not the all-you-can-eat place?'

She shrugs. 'Probably. I'll go and get Afia's bottle ready.'

I sit with Jock until he pokes me and mooches away, dejected, back into den two.

Romina and Afia have moved into den three and sit on a platform above the glass that spans the public area. Romina has made a nest of wood-wool and I can see her feet poking out over the edge, lying on her back. She must be asleep as she hasn't stopped Afia dangling from the platform from one arm and patting her chest in Kera's direction.

Kera sits on the glass above the public, with her back to the wall and watches Afia intently.

Afia continues to swing and shows her grin play face.

Kera glances into den two. Jock is curled on the floor, facing the wall. The rest of the group are outside.

Afia drops from the platform with a thump.

Kera grunts.

Afia takes a step toward her.

Kera casts another glance about the enclosure and her eyes flick between Afia and the sleeping Romina up on the platform.

I move down the line of feeding dens to get a better view.

Afia blunders toward Kera, trips and rolls on her back.

Kera reaches out an arm and Afia stretches to bat at her fingers.

I hold my breath.

Hannah creeps down the corridor to join me. 'Saw them on the monitor,' she whispers.

Kera hooks Afia toward her, slides her across the glass and rubs her belly.

Afia's face stretches into a grin and she play-bites at Kera's fingers.

Kera bursts upright, her body infused with sudden strength and energy. She cuffs Afia, sends her across the smooth glass on her back.

Afia bounds to her feet and runs back at her.

Kera knocks her onto her back and spins Afia the length of the public area and follows up by doing a pirouette, twirls on one heel and kicks out her other leg. Kera normally claps when she performs this manoeuvre and only does it in the mornings, when she gets worked up, at about 8.45 a.m., brought on by the anticipation of breakfast.

Afia doesn't know what to do and sits still on the glass, Kera in between her and Romina.

Kera clutches her own ears and shakes her head. Waggles her lips and then sits down again, remains still and grunts quietly at Afia.

'She's pretty wired,' Hannah murmurs.

We hold our nerve. Let the interaction play out and long for Kera to get it right.

Afia stands up.

Kera grunts.

'She's not clapped on purpose,' I say. 'Doesn't want to wake up Jock or Romina.'

Afia falters as she trundles across the glass toward her.

Kera lunges forward and takes an arm in one hand and Afia's foot in the other. Bursts upright again and spins Afia as she pirouettes.

We see the play grin on Afia's face fix in panic.

Kera whizzes her up and down as she spins, arms at full stretch.

'No, Kera,' we both shout.

Romina sits up in her nest.

Kera releases Afia, as the waltzer spin brings her in line with the glass. She cartwheels and rolls across the glass in a tumble of thuds.

Romina leaps from the platform and coughs.

Kera goes full self-applause, a series of claps between her cupped palms, and continues to pirouette.

Afia runs full pelt back to Romina. A speed I've not seen from her before, a crouching run, arms going like a set of manic crutches. Her lips wobble, but she doesn't hoot, casts a nervous glance back at Kera once she's safe in Romina's arms.

'Kera.' Hannah shakes her head. 'Ever heard of playing gently?'

Kera spins, stamps and claps, shakes her head and then scarpers as Jock comes charging up the ladder.

If Kera had let go of Afia and she'd hit the wall, at that speed, it could easily have been it. Unknowing infanticide. There was no sign of any maternal bond. Just the fact that someone wanted to play with her seemed enough to tip her over the edge into manic delight. Kera showed she had no understanding of babies and how to interact with them, having never learned how to moderate her own strength, exiled as she was from gorilla society. Afia stayed clear of Kera from then on, just like everybody else.

Chapter 14

CT Scan

Almost a year after Sal's death, the group dynamics had settled. Trim Touni was top female, a position Romina was happy for her to take on. Another female gorilla joined the group from a zoo in Germany, as part of the European breeding program. Her name was Kala and genetically she was a good match for Jock.

Kukena, Jock's eldest daughter, had almost come through the annoying, pissing-everyone-off phase. Without her mum Sal around she would choose to sit with Romina if she was feeling down and had begun to play with Afia a bit.

Afia and Touni's daughter Ayana were inseparable and spent as much of the day as possible wrestling, usually in the proximity of Jock. The pair of them had learned that Kukena and Kera would leave them alone if they hung around him and he would occasionally throw out one of his tree-trunk woolly arms for them to pat at or climb over.

Kera, as ever, was on the periphery.

Kala, the new arrival, has short bandy legs and a long, solemn face. Her crest is similar in style to Romina's, a neat fan off the back of her head. Kala's natal group in Germany was less established than Touni's had been and there is a marked difference in Kala's behaviour on arrival. Whereas Touni had read the group and successfully figured out a way to not only join but go on to then manipulate everyone to her advantage, Kala has less experience of group dynamics and so for a while, she languishes at the bottom of the hierarchy, failing to make friends with anyone. Once Jock mates her, however, she overtakes Kera. The group is calm, which is generally a sign that everything is about to kick off.

I rarely work with the gorillas anymore, but I still give Afia the odd bottle feed through the mesh. At this point Afia is receiving three milk feeds a day.

As I am still running the small mammal team, I miss Romina having a knockdown. She has become increasingly lethargic and begins struggling to pass faeces, however her knockdown, putting her under general anaesthetic so she could be examined by the vet team, doesn't reveal anything conclusive.

During the procedure Afia has been left alone with the group. She has to run the Kukena gauntlet a few times and is let back in with Romina as soon as the vets have exited. She clings to her, sucking her thumb. A bleary Romina hugs her close as she slowly comes around from the anaesthetic.

The inconclusive result means we need to give Romina a full CT scan as soon as possible.

The first available date for the mobile CT unit is slap-bang at the beginning of the Christmas rota, meaning our operating numbers are at bare minimum. Not just the mammal team, but the other departments too. Maintenance and Michelle the head vet are away on Christmas holidays.

The CT scanner sits at one end of a long trailer. A low-loader lorry which functions as a full CT unit. The sides of the trailer, once it is parked, extend outward to form the observation office, accessed by a set of metal steps and a hydraulic lift to allow patients to be wheeled to the unit and lifted inside.

The truck is enormous, and I know from years ago, when we had to scan some lion cubs, that this giant vehicle only just fits through the access gates. The only way it can get on site is to park up on the top terrace by the lion enclosure and gift shop, on the opposite side of the zoo to the gorilla house. It has to be backed through the two sets of double gates, with centimetres to spare on either side, and the trailer has to fit all the way in so it can open up and extend once within the zoo walls. As the cab part of the truck is so long, it is only possible to close one set of double gates, meaning a single barrier between the zoo and Bristol.

This is against procedure in terms of our zoo license. Dangerous animals must be contained by dual gates, so that should they escape their enclosure, they should remain contained within the zoo grounds. As one of the sets of gates has to remain open, mitigations have to be put in place, particularly as we are taking Romina, a fully grown gorilla, out of her enclosure, wheeling her across the zoo and into the CT scanner.

The scan is scheduled to begin after the zoo has emptied of public for the day and due to the Christmas rota, it means I have to provide gun-crew cover, the first and only time I take a real gun out of the cabinet. I deactivate the alarm and unlock the gun room. Fill out the incident book. Add the time and date, my name, which weapon, number of rounds taken and the reason. *Taking gorilla out of enclosure.*

I pick the .308 rifle from the cabinet and check the breach is empty. Take the firing bolt off the shelf and slide it into the weapon. Make sure the safety catch is on and double check it. The two magazines each contain five hollow-tipped bullets, slotted in one above the other. I place both magazines in my pocket. Check the safety catch again and dither over the suppressor, the long screw-on silencer. Without the suppressor, firing indoors will burst eardrums, but if I screw it onto the end of the barrel, the rifle is much longer and unwieldy. If the worst comes to the worst and I have to shoot Romina, kill this gentle animal who has become a parent to Afia, a loved and loyal mum, if she were to wake up and attack someone in the back of the CT trailer or as we transport her out on a stretcher, then I need to be as unimpeded as possible.

I was interviewed before joining the gun-crew team. One of the questions is of course: could you kill one of your animals?

The answer is: I don't know, but I'd said yes. If it comes to a life-or-death situation I'd have to, but how can you know at the time, in the thick of it, with an animal you've worked with for years and grown to love? How do you know you would actually pull the trigger and if you did how the fuck would you ever get over it?

We fire live ammunition when we practise at the police firing range. The targets are timed to twist and face you for 10 seconds, which sounds like plenty of time, but generally isn't. Hitting a moving target is hard and firing a rifle within the zoo grounds is fraught with danger. The speed that the bullet exits the gun is somewhere in the region of 18,500 miles per hour and could easily kill anyone unlucky enough to be in its path. If I fired and missed, and the bullet flew over the perimeter wall, it would travel 4 miles across the city. The zoo is split into safe fire zones and I run through them in my head, directions where I can guarantee a backstop of the perimeter wall or buildings if I miss.

I take the suppressor and slide the rifle into its slip. 'Slip' is the gun-crew term for the zip-up carry bag. Pull on my hi-vis bib and collect a set of ear defenders. I know the chance that Romina will wake up and go berserk is practically non-existent. I hold onto that thought. I'm following procedure in relation to a vastly unlikely event.

The vet team join us at Gorillas.

Rowena is leading the procedure, as Michelle is away. She holds up the risk assessment. 'This is the reddest one of these I've ever seen,' she tells us cheerfully as we all gaggle together to sign it. Rowena has been a vet at the zoo longer than I've been a keeper. Luckily for us and the gorillas, we all trust her in exactly the same way we do Michelle and Teresa.

'This is potentially very dangerous,' she informs us.

I hold my breath to listen.

'We'll be taking an adult gorilla out of its enclosure.'

I nod along with the rest of the team.

'We're going to wheel Romina on the little truck thing.' Rowena pauses.

'The dilly,' Bob supplies from behind me.

She nods and smiles at Bob. 'Teresa will monitor,' she continues, 'but until we get Romina into the scanner, she won't be on any gas.'

Bob mutters something behind me that I miss.

'In the event she wakes up—' Rowena pauses '—and I must emphasize that is extremely unlikely, but if she were to wake up, follow my instructions. Move away to a place of safety. Do not get in front of Al, or myself.'

Hannah glances at the gun, still zipped up in its case.

'Al will have the rifle, which will be loaded with live ammunition, and I'll have the dart gun prepared. We don't want anyone getting shot accidentally.'

We all nod.

'Hannah,' Rowena says. 'Do you want to fill everyone in on the animal side?'

'Thanks, Rowena.' Hannah's cheeks look flushed in the face of her solemn audience. 'Bob's going to inject her by hand.'

He shuffles behind me as everyone looks his way.

'We're going to set up the back half of the isolation den as a retreat space for Afia,' Hannah says. 'In case anyone tries to pick on her, shut the slide on a break, so only she can squeeze through. Too small for Kukena and Kera.'

I glance up at the monitor on the wall. Romina is already separated and Afia clings to the dividing mesh, baffled. I can see by her face, her pouting lips, that she's hooting. They should be snuggling up together to sleep by now, but instead she can sense we're all still here as can the rest of the group. Prowling the house.

'Once Romina comes back,' Hannah continues, 'we'll put her in the front half of the isolation den and give Afia access to her. We've had to override the light timer in the corridor. Bob, make sure you turn them off at the end of play.'

'Will do,' he says.

We all wait in case Hannah wants to add anything else.

'Right,' Rowena says. 'Good luck everyone.'

I wait outside the gorilla house with the gun. I obsessively check the safety button, the magazines still in my pocket. I've screwed on the suppressor, not that it means the shot will be silent. When you fire with a suppressor it makes a dull thump, rather than a massive bang, but you can still hear it, even through ear-defenders. I run through the route in my mind, visualize the interpretation boards, the railings outside Monkey Jungle. Places where I can rest the rifle barrel, to keep it as steady as possible. Feel my skin prickle.

The gorillas burst into shrieks inside.

They're bringing Romina out.

I check the safety button is on one more time and take a magazine out of my pocket. Slot it into the gun.

The door opens and Rowena leads the team outside. I fight the urge to help, remain to one side as Bob and Hannah, Teresa and Twilight Scott stagger outside with Romina between them on the stretcher. She is piled with blankets. A crackling space blanket wrapped on top.

Teresa holds Romina's hand and we're off. A creaking procession through the wet, dark zoo. Wellies slap at the backs of legs. The wheels of the dilly creak. Bob pulls it along with Hannah, Scott holding the electric blanket over

Romina and keeping the stretcher handles from snagging in the wheels.

I keep pace, the gun pointed upward.

The spider monkeys go berserk as we trundle through Monkey Jungle, calling out in fear. The ring-tails join in next door.

'So far so good,' Teresa says as Bob and Hannah drag the dilly past the restaurant.

A couple of late kitchen staff gape as we roll by.

The rain drenches us, pinging on Romina's space blanket. The trees along the top terrace sway in the wind and I see the trailer lights next to the lion enclosure.

No one talks, just puffs and pants and the dilly wheels squeak.

The hydraulic lift is down and ready. The scan team wait above, sheltered in the office, all dressed in vet scrubs.

Bob and Hannah wheel Romina onto the lift plate, Teresa with them, and the operator guy clangs shut the safety rail. Armed with a remote control not dissimilar to the one we use in Gorillas, he lifts them all up and into the waiting trailer.

I climb the set of metal stairs and wait as the team lift Romina off the dilly and onto the table, more like a bed. The bed that will eventually deposit her into the curved tunnel of the scanner. The vets and scan team work in their usual efficient manner. Quick and precise. Conversation is kept to a minimum until the breathing tubes are inserted down Romina's throat and she's stabilized under anaesthetic.

The rain continues to batter the trailer roof. The office inside is cramped and now Romina is under, there is no way she'll wake up and cause havoc. I need to wait outside to give everyone room.

I take cover, sheltered from the rain by the covered walkway that runs the length of the lion enclosure. The lions have disappeared. The rain and wind have put them off the light show outside and they've both gone back to bed.

I check the breech of the gun is empty and remove the magazine. Put it back in my pocket and zip the gun back up in its bag again. It's going to be hours before Romina comes out again and we whisk her back to the enclosure.

I stamp my feet and wonder what the time is. Sparky is at a Christmas do tonight in town. I'd be there with her if I wasn't standing around like a sentry.

The trailer groans and rumbles. I see the odd head pass by the office window.

I wait and shuffle to keep warm.

'Marvellous bit of kit.'

I jump.

The operator guy is back again. 'When you really think about it,' he says and nods at the trailer. His breath comes out in clouds.

'It must take ages to learn how to set it all up?' I say.

'It's incredible.' He nods.

'Is it true the magnetic force is strong enough to mess up people's tattoos?'

He gives me a funny look.

'I heard that somewhere,' I say. 'The black pigments in tattoos get dragged through the body.'

'We had a liquid hydrogen leak once,' he offers. 'It came out of the pipes at this titanic pressure and totally engulfed a tree, froze it like a scene from Narnia.'

'You what?'

'What's this one in there?' he asks. 'A chimpanzee?'

'A gorilla,' I reply. 'She's the surrogate mum to the one that's been in the news. The baby we hand-reared.'

'Oh right,' he says.

'She's an amazing animal. Took on Afia as her own baby.'

'Yes,' he says, with sudden enthusiasm. 'I saw it on the news.'

The door at the top of the steps leading up to the trailer opens and Bob pokes his head out.

'Al,' he says. 'Might want to come in.'

I climb the steps into the trailer.

Hannah has tears in her eyes.

'What's up?'

Rowena gives me a sad smile. 'Unfortunately we've found a large mass in her abdomen.'

Romina lies on her back, the tubes in her mouth. The office is full of screens, images of curled bright colours. The scan team whisper and point and fall silent as we gather around the bed. Her huge arms lie on either side of her furry body, head supported on pillows. Her chest rises and falls, kept steady by the anaesthetic machine.

'It's inoperable,' Rowena adds gently. 'It will be kinder to put her to sleep now, rather than let her struggle on.'

'What about Afia?' I blurt out.

'Romina will be in a considerable amount of pain, any relief we give her will have to be in very high doses and it really isn't fair. We're talking days or weeks.' Rowena's face is lined with genuine concern and sympathy, flat and down like the rest of us. 'I'm sorry.'

I look to Teresa, the vet nurse. 'It really is the best thing we can do,' she says. 'She won't suffer or feel anything.'

I know they're right. Trust them both implicitly. I look at Romina, unconscious and unsuspecting, laid out on the bed between us.

We gather around and I take her long hand in mind, the fingers dry and cool to the touch. I put my other arm around Bob. He pulls Hannah close, the three of us next to Romina as the vets work on administering the drug that will end her life.

My eyes swim with tears.

Afia is going to be devastated. Romina was so great with her, the two of them loved each other, the gentle play, nesting together. Romina laughing as Afia wrestled her, holding her close when she was scared. Hanging out together outside, watching over each other, a genuine relationship of maternal care and all that was about to come to an end.

We cry and sniff as Romina slowly drifts away.

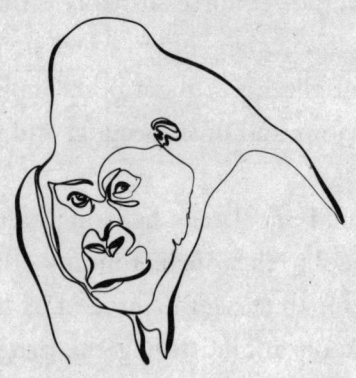

Part Three
Single Parent

Touni
Perfect mum

Jock
Still the Silverback,
but older and less
prone to violence

Kala
Hopeless mum

Ayana
Bit of a menace

Perfect Juni
The favoured
one

Hasani
Rejected by
his mum

Kera
Eldest and
biggest female

Afia
Juvenile

Chapter 15

Just Before the World Went to Shit

Romina's death meant Afia spent a few lonesome nights, tiny and alone, sleeping on a ledge above one of the slides. It was a heartbreaking turn of the screw on top of losing Romina, but by the New Year Afia had learned to hang around with Jock for security. Her granddad, silver furred and huge, kept her safe from everybody else and she stuck with him for the next couple of years. We had stopped bottle feeding Afia by this point as she was now four years old and completely imbedded in the group. The final stage of hand-rearing her was over.

Afia and Ayana, Touni's young daughter, were both growing up strong, their wrestling occasionally getting out of hand. They were getting into the lanky, long phase of their growth, with powerful muscled arms and tons of energy.

It had been months since I'd worked with the gorillas and as much as I was still attached to Afia, the relationship ebbed away, which in the professional side of my mind I knew was a good thing.

Jock was slowing down, approaching his latter years, and spent more time sleeping. He'd still occasionally launch at Kera but was less likely to cause horrific injury.

As ever Kera remained a social outcast. Her hair-plucking behaviour got worse and she gave herself bald arms and visible patches of dry, grey skin across her belly. Our enrichment efforts went into overdrive to keep Kera occupied and engaged, but her fundamental problem was desperate loneliness. She was too rough to play with her unknown daughter Afia or Ayana, and if Kera tried, Jock would jolt out of his slumber, murderous at being awoken. In his bleary state and fuelled by his habitual zero tolerance of Kera, he would leap to his feet, assuming his beloved offspring were under attack. Kera was able to evade him almost anywhere in the enclosure these days. We had added extra ropes and webbing, high above the enclosure floor to make sure Kera always had an escape route and she would scurry off to sit casually on a rope high above his head and wait for his rage to ebb away.

We'd initially hoped that Kukena, Jock and the departed Sal's daughter, would form a friendship with Kera when her mum died, but she too soon snubbed her. Kukena continued to be a menace to society and spent her time trying to get the others beaten up by Jock. Her position within the hierarchy wavered and tragically, she performed the worst act of malicious behaviour I've seen in all my years as a zookeeper.

Trim Touni remained as head female and ruled with a swagger of assumed position. She had Ayana, her daughter, to back her up and to a degree Kukena too, who she befriended when she first arrived all those years ago, to secure Dangerous

Sal's protection. Touni's face was always a solid frown and she mainly mistrusted us, with no real willingness to engage.

Which leaves Kala.

Kala loves to play and to begin with has a touch of the Kera about her, in that occasionally she still misjudges the mood, never quite sure what she's done to upset everyone. She vies with Kera to not be at the bottom of the pecking order and secures her position when she gives birth in the summer, but proves to be an inexperienced mum, enabling Kukena to grab her two-week-old infant and run off with it.

None of us are at work when it happens and we trawl back through the CCTV footage for clues when we come into work to find the baby looking weak and unresponsive.

The footage shows Kukena snatch Kala's baby away.

Kala gives chase and Kukena, clutching the baby by the ankles, smashes it against the brick wall in den one. Every so often horrific things happen between animals. As a zookeeper you begin to build a wall to cope with the death of your animals. Like Bob, I have to find a way of coping with the reality of zookeeping, but I really wish to this day that I'd not seen the footage. A week later, Kala carries a floppy little corpse around with her, piles wood-wool on top of it when it begins to smell.

Life rolls on.

I was now the permanent team leader of Mammals North, which includes the nocturnal house and Asiatic lions, tree kangaroos and meerkats. I am busy and like the rest of the world, have no idea what is coming around the corner.

Life rolls on for the gorillas too.

Kukena joins a family group in Spain as she reaches breeding age. Kala is due to give birth for the second time in August and Touni has just conceived when the zoo closes its gates to the public.

We enter lockdown against Covid-19.

Primates need to be with their own kind to thrive. I can only imagine the misery of going through lockdown alone, in solitary confinement, with limited access to the outdoors or any chance to socialize face-to-face at all.

I am aware that I'm speaking from a privileged point of view here. I'm lucky enough to live in a society with a distinct lack of physical violence and relative wealth. As a male, living in a city in England, I enjoy a level of freedom not experienced by a vast proportion of the human species across the world, but regardless of who we all are, life as we know it is about to change.

As the pandemic rolls around the globe, and our movements are restricted, our habitat shrinks enormously. We no longer have access to our usual haunts, the regular territory we exist in.

The monkeys and apes at the zoo continue to live as normal, in their social groups together, but for us, as highly social primates, we are about to experience life in captivity, at home.

For me, as a zookeeper, lockdown is very different.

The vet department leads on our coordinated response. Our volunteer teams, from home, sew us face masks to wear as we work, with pockets to hold filters. Before the market catches up, we look like bandits, taking all precautions not to pass anything on to the animals we work with. There is a report early on that some lions have caught Covid at a zoo in the States somewhere. The red pandas and giant otters are deemed particularly vulnerable and so too of course are the gorillas.

The fear of passing Covid on to endangered species, knowing that you could be responsible for their death, weighs heavily on us all, an added layer on top of the existing concern for elderly and vulnerable family members and colleagues. We are wired, twitchy and taking it all seriously.

As gorilla keepers we are issued with half-face masks, which have a filter in a nose cone and tight rubber seals that clamp over the mouth and nose. We develop pus-filled spots and rashes around our mouths as we sweat and continue to work in the sudden stifling heat.

Anyone classed as vulnerable works from home. Non-essential staff are furloughed, including the gardens department, and as life outside becomes more constrictive, the zoo begins to rewild itself. With the hot sunshine and no one to cut back and prune, the trees and plants explode in rampant growth and the pathways have become overgrown, heavy with shady canopy cover. The grass on the main lawn grows long and soft and we all gather there at lunchtimes, spaced out in a socially distanced circle, plagued by flocks of jackdaws, which have suddenly become deprived of their usual diet, the crumbs and cold chips that zoo visitors could be relied upon to leave behind.

Covid doesn't hit the team immediately, but as vulnerable people are shielding, our numbers reduce to the bare minimum. When I am on site, I have become the most senior member of the animal team. Higher-ups are reachable by phone, but the day-to-day running of the zoo is down to me and the rota becomes all-consuming. Managing food deliveries and keeping my fingers crossed that nothing breaks on a section I know zilch about: the aquarium or Bug World, reptiles or birds. Life has an all-hands-to-the-pump atmosphere. The team pulls together, as usual in a time of crisis, but as the pandemic grinds on, they begin to wobble. Hearing the death toll each day, steadily rising and with no firm talk of vaccines, we all begin to struggle.

No public means no talks or engagement.

The gorillas don't seem to notice the lack of people but have become more reactive to distant noises that would usually be drowned out by general hubbub. The group spends more time without human contact as we have shortened our working days by a couple of hours. It means more time for us shut in at home.

Cooking has become a time of peace for me, locked in as I am, supported by Radio 6, a drift into the here and now, distracted from the pandemic chaos. I'm halfway through cooking a Greek bean recipe when my phone goes.

I glance at the clock on the cooker and know it will be Hannah, back home by now, unless something's kicked off and she's been called back in.

'Hey,' I say. 'Everything okay?'

'Hi,' she responds and I can hear her tone is off. 'Yeah, nothing to report for tomorrow, we're still trying to get a weight off Kala.'

'Baby bump is looking big,' I say. 'We're about six weeks off due date, aren't we?'

'August,' Hannah replies. 'Are you busy?'

'Nope. Cooking, what's up?'

Hannah pauses. 'I just needed to talk to someone. Well, I needed to talk to you really.'

I fish my headphones out of my pocket. 'Are you okay?'

'How far through your one-to-ones are you this month? With your team?'

'Not far.' I connect my headphones and push the buds into my ears, so my hands are free to dip a teaspoon into the sauce and blow on it. Every month Hannah and I are meant to sit down with each of our team members and plan their work objectives going forward. 'I know what you're going to say.'

'I don't have a degree in counselling,' she says.

I can taste the cinnamon and oregano, as promised in the recipe. 'I know. I'm feeling dangerously underqualified.'

'They're all depressed, anxiety-ridden, and I don't know what to say to any of them.'

'Same for me.' I add some stock from the saucepan. 'A couple of mine are okay, it's not the whole team, but yeah. It's tough. I keep rambling on about primate behaviour.'

'Really,' she says. 'You surprise me.'

'Our territory has shrunk, the usual habitat, no wonder we're all stressed.'

'Like locking all the gorillas into the isolation den.'

'Exactly. Most of our normal habitat is being denied us. Pubs, can you imagine going to a restaurant again? Shut indoors with loads of people.'

'I can imagine it,' she says.

'How's Bob doing?'

'How do you think?' Hannah sighs. 'He hates change.'

'And he's on his own. Isolated.'

'He'll be stereotyping in no time,' Hannah adds, but the truth stops it being funny.

'Stereotyping' in zookeeper speak refers to the repetitive actions captive animals take on when they're stressed.

'How are you doing?' she says. 'Has anyone asked, or do they just vomit their depression all over you?'

I pause at the word vomit, spoon raised to taste the sauce again. 'I'm alright. I've turned the dining room into a workshop, making these shelves for Sam's twenty-fifth birthday.'

The sauce tastes bloody great.

'I thought Sam lived in Belfast?'

'Fuck knows how I'll get them delivered,' I continue. 'Doing something tactile is keeping me sane. It's fun, you know, satisfying.'

'Proves enrichment works,' Hannah says, referring to the sheets and duvet covers we give the gorillas to play with, the puzzle feeders and cardboard boxes, things to keep them entertained and using their brains.

'I've got a keeper who provides it to me almost daily,' I say.

'A keeper?'

'A delivery person.'

'Bet Sparky loves that,' Hannah says.

'She's been great about it.' I turn off the beans and let them sit while I get on with the potatoes. 'Kind of resigned to it I guess. I've got shit everywhere and I've drilled two holes in the table.'

Hannah laughs. 'Idiot.'

'I don't know how Sparky manages all the Zoom stuff. You and me socialize with other actual people at work. For her it's all screens, or me cooking and drilling holes.'

'How are the kids coping?' Hannah asks. 'I mean, I know they're both grown-ups.'

'They're okay. Fizz has a flat across town, it's meant to be the last year of uni and all the classes moved online. We've got the family catch-up on Zoom later.'

'I'll see you in the morning then,' Hannah says. 'Thanks for letting me vent for five minutes.'

***

The zoo remains closed, and no public means no income.

A ripple of concern is raised at an all-staff meeting, conducted on Zoom, complete with a dodgy screen backdrop, presented to a host of switched-off cameras, all muted. There is talk of honesty and transparency. No idea of when we'll reopen again to the public and what that means in terms of finances.

We continue to work, sweating in our masks, testing daily, and eventually Covid restrictions begin to lift.

The rampant foliage is cut back and fed to the animals. The main lawn is mowed, and the paths covered in arrows, designating direction of travel. One-way systems are constructed, with certain areas deemed off limits.

There is a slow fragile sense of optimism; life hasn't returned to normal, but there is a loosening of rules and restrictions and people are desperate to get out. The public coming back to the zoo means an income stream again and there is talk that our insurance might pay out, a hope that miraculously, the zoo has cover that could potentially be applied to the dramatic loss of income from Covid.

My team remains stressed, but it feels as if there is a slight glimmer of hope. Most of my species remain off-show to the public. It's hard to think of anywhere more difficult to maintain social distancing than a nocturnal house in a zoo. Twilight World is made up of multiple enclosures, all in the dark. People bump into one another and sneeze on the glass windows, clutch handrails, paw over dimly lit interpretation signs and mill about trying to catch sight of the animals. No amount of barriers and arrows spray-painted onto the floor is going to offset the risks of reopening it to the public.

We don't yet know that it will never open to the public again.

Kala gives birth in August. The infant, named Hasani, looks strong and the rest of the group are respectful and don't try to interfere. Kala lets him feed to begin with, never for very long, but in comparison with her last baby she is doing much better.

He gets to four weeks old before we step in to save his life.

As with her previous offspring, Kala begins to abandon Hasani and seemingly forgets he is there, leaving him behind to get dangerously cold. He is expending more energy than he is receiving in calories and Kala also begins limiting his feeds.

Attempts are made at supplement feeding, but it becomes clear we either have to euthanize him on welfare grounds, rather than watch him starve to death, or hand-rear him, with the aim of getting him back in with Kala once he is strong enough.

We go for it.

Chapter 16

Closing Time

There is a flat on site at the zoo. Before lockdown it had functioned as a mini hotel getaway. A package deal where up to six people could come and stay on site at the zoo overnight. Have a meal prepared for them by chefs in the flat kitchen, while they got a private tour of the zoo in the evening. It has also been keeper accommodation over the years. An upstairs flat with a tiny balcony that overlooks the tapir enclosure.

Due to Covid-19, the flat is no longer functioning as a mini hotel getaway. It is free, and rather than expose Hasani and by extension us and the rest of the gorillas to an increased risk of infection, it is decided he will move in. We figure out the rota for hand-rearing and it breaks down to staying three or four nights a week on site, interspersed with taking the following day off to recover as he is at the age where he requires milk feeds every 2 hours. This time around we are doing it as single parents. Isolated and alone.

I join Bob and Hannah in the messroom. We sit away from each other in the corners to maintain as much space as we can.

'Here Al,' Bob says. 'What did you think of that all-staff meeting?'

'I couldn't get my phone to connect,' I say.

Bob is tense, shoulders hunched, and his face is red above his mask. He has tears in his eyes.

'I missed the lot,' I add.

'Closing time.' He shakes his head. 'Can't fucking believe it. Selling the zoo and all the money going to Wildplace.'

'What?'

'We'll all lose our fucking jobs.'

'Not necessarily.' Hannah holds up a pacifying hand. 'That's not what they said, Al.'

I slump down into a seat at the end of the table. 'Wait what? We're closing down?' I feel cold and my toes tingle. One of us generally has to miss the all-staff meetings, as we can't leave the gorillas on their own, especially not now. 'It's all going to Wildplace?'

They both nod.

Wildplace is Bristol Zoo's sister site, an area outside the city with more room and plans to expand.

'Some of us will lose our jobs, Hannah,' Bob snaps. 'I've been here since I was seventeen,' he adds and his voice cracks. 'Been coming here since I was a kid.' He sniffs behind his mask.

'Don't, Bob,' Hannah says and waves her hands in a socially distanced hug. 'We don't know any of that yet. These guys will be moving to the new site.' She points at the monitor on the wall.

The gorillas are all eating their tea. Kala has the whole of den three, Hasani asleep on the floor next to her. The rest of the group are all shut into their individual feeding dens as usual.

'We'll find out more in a few months,' Hannah says to me and raises her eyebrows, nods her head at Bob.

He's got his head in his hands at the end of the table.

'The Higher-ups are working on a species plan,' she adds. 'There'll be a brand-new gorilla house. We'll all be involved with the design.'

Bob snorts. 'Yeah right.'

'We will.'

I glance out of the window at the seal and penguin enclosure. 'What else is moving out there?' I say. 'What's happening to all the animals?'

'That's what the Higher-ups are doing,' Hannah explains. 'With the species plan, figuring out what to take and what will need to be rehomed.'

'Did they say anything about a nocturnal house?'

She shakes her head. 'You can watch it later,' she says. 'The meeting was recorded.'

'Did they say when?' I ask.

'Didn't tell us fuck all,' Bob growls.

'Come on, Bob.' Hannah sighs. 'Watch the meeting, Al. They said they're going to come back to us in a few months' time. It's top secret too, they don't want it getting out to the press yet, not until they've finalized stuff.'

We all stare at the table.

'Are we doing this?' Bob says and lifts his eyes to the monitor.

The gorillas have finished their tea and wait to be let out of their feeding dens and mixed back together.

I take up my position in the upstairs keeper-corridor and sit unobtrusively at the end feeding den.

Closing down?

The thought engulfs me, yet another undercurrent of uncertainty in the current whirlpool of life, flailing about in a washing machine of bullshit.

If the nocturnal house closes, most of the species I look after will have to be moved on. Wildplace already has some big cats, the cheetahs, and everyone hates the Asiatic lion enclosure, so it's unlikely they'll be moving to the new site. My job as it stands is going to end, but we've got an immediate focus right now.

Hasani. We're about to pull him out of the group to save his life. Despite the world being in chaos, we're going to hand-rear a second baby gorilla. That we can do.

'Opening the slides,' Hannah calls and begins to work the remote control. The hydraulics groan and tick and the slides clunk and grind open.

The group are all let out of their feeding dens and mixed back together in dens one and two. Everyone except Kala and Hasani.

Kala is desperate to be with them, can't bear to be separated and scurries straight into the line of upstairs feeding dens, set up to create one long row.

We shut her in.

Hasani hangs from her belly and his head droops from his shoulders, big eyes closed in his pale face, the skin stretched with wrinkled dehydration, his tiny body a sack of prominent bones. He trails one foot as Kala stalks the feeding dens, hopeful that we'll let her out with the others.

Hannah continues to work the remote and the rest of the group mill about at the far end of the row. Kala rushes to be near them, despite still being separated by the mesh.

'Kala,' I call and shake a tub of pellets to bring her back.

The rattle causes Jock to amble over. He can't get to me and the pellets as the row of feeding dens and Kala are between us, but he grumbles and stoops to peer through the mesh.

I pull the lid off and push some monkey pellets into the end feeding den. The tasty stuff.

Jock grumbles and sits as close as he can get and Kala stops to eat, happy to be in his vicinity, even though the slides remain closed between them.

Hasani's dangled foot is clenched tight, the instinct to grasp keeping the toes curled in a useless fist. His head nudges at Kala's chest and his eyes flick open. The bony arms heave his tiny frame, and he tries to latch on to Kala's nipple.

She brushes him off.

He cries and works his mouth.

She bares her teeth and tugs him from her belly. Disentangles his fingers and puts him down. Drops a pile of wood-wool on his head.

I give Hannah the thumbs up.

'Kala,' she calls and shakes a tub of peanuts.

The gorillas all know the difference between rattling peanuts and monkey pellets.

'Come on, darling,' Hannah calls.

I hear the patter of nuts falling through the mesh, up the row of feeding dens.

Jock rushes away from Kala at the sound and she leaps along the row of feeding dens to keep pace with him and abandons Hasani.

'Close it,' I call.

The slide into the end den, where Hasani lies motionless under a pile of wood-wool, grinds closed.

I open the keeper-door and step in with him.

Kala freezes, spins to look at me as I pick up her baby. Cuffs her head and races toward me. Forward rolls and cranes her neck as I step back out of the enclosure, Hasani cradled in one arm.

I lock the keeper-door and Kala screams and shrieks, keeps pace with me as I carry Hasani up the corridor, his body cold and floppy. I head straight out into the messroom as Hannah opens up the rest of the slides to let Kala back with the group.

'Nice one,' Bob says. His misery at the all-staff meeting and announced closure of the zoo has been put on hold and pushed away. 'I'll radio the vets.'

I place the limp body in a nest of vet bed, with blanket-wrapped hot water bottles. His skin is papery and tents across his concave shoulders.

Hannah joins us in the messroom and pulls off her rubber gas mask. She replaces it with the cloth masks we are to wear around Hasani and looks up at the monitor. 'Look how stressed Kala is.'

'Must have mastitis, I reckon?' Bob suggests. 'Why she won't let him feed.'

I see Kala on the screen, watch her scurry across the glass above the public area and do two more forward rolls, one after the other.

'Is that your original string vest?' Hannah asks me.

'Might be,' I say. 'It was up at the vets on the hand-rearing shelf, with all the bottles and sterilizer.'

'Have we got all the kit?' Bob begins to paw through the bag.

'I've checked it,' Hannah says. 'Thank god we don't have to sleep in these gas masks. Can you imagine? There's more filters for our masks in there. The slide-in ones. Get changed every twenty-four hours.'

Bob rummages through the bag. 'Need to get some teething rings,' he murmurs.

'I'll give you a hand over to the flat,' Hannah suggests. 'After the vets have seen him and he's had his fluids. Get you both settled in.'

'It's plush over there,' Bob states. 'Went for a nose around earlier.'

'Not into my room I hope,' Hannah says.

Hasani opens his eyes, tucked away in the nest we've made him.

'Had to scent mark everything.' Bob forces a laugh. 'Anyway, never mind that. I'll chuck a seed scatter through the roof.' He nods at the monitor. 'Distract Kala, bung a load of enrichment in, poor cow, look at her.'

Kala still stalks the house.

'Al,' Bob says. 'When you've watched the all-staff later, give me a bell.'

'Okay,' I reply.

The first night with Hasani is horrific.

He was abducted, taken by weird hairless bipedal apes. Different smell and an environment beyond all recognition. We give him fluids, injected under his skin, and I hold him as he whimpers and clutches at me in pain, with his long, spidery fingers. His hair wispy and sparse, face pulled in fear and misery.

The first feeds with Afia had been intense, with the constant fear of aspirating her, before she got the hang of suckling from the bottle. Hasani is different. He's fed from Kala, when she would let him, and received all the colostrum, the first milk produced, full of antibodies and high in vitamins. If it wasn't for Covid-19, we'd be more relaxed about his resilience being around us, way more so than we were with Afia in the early days, but she had been premature and vulnerable. The realization that we didn't even wear masks with Afia seems beyond incompetent now.

His suckle response is better. He's starving, and despite being as controlled as possible, I get a ribbon of milk from his nose and know I'll have to nebulize him in the morning, subject him to the giant gas mask across his face to ward off any risk of pneumonia.

I wind him and grumble. The first feed gives him terrible cramps, as the change in milk from gorilla to powdered formula makes him writhe in pain. He cries and snivels, his skinny body weak and feeble. Afia had been thin too when she was born, her spine a line of knobbles, but Hasani is skin and bone. His triangular shoulder blades and delicate ribs are all wracked with pain as his body tries to digest the milk. I use every trick in the book, gently manipulating his legs, rubbing his back and belly. His pain is tempered with bewildered fear, having been ripped away from everything he has known. We walk the flat. Two main bedrooms, two small bathrooms.

'I threw up in that bath once,' I tell him, keeping a soothing tone, holding him against my string vest. 'Not this actual bath, the one that was here before the place got revamped.'

I put my head into Hannah's room. It has a long mirror on a stand in the corner. Save showing him his reflection for later down the line. 'This place was keeper accommodation once,' I say. 'We had a party here and Bob made us all sick with his home brew.'

Hasani hoots, low rather than scream pitch.

I take him into the kitchen, sealed behind heavy fire doors. The cooker is new, the fridge near empty, except for my dinner, cold pasta in a plastic tub, ready to be nuked in the microwave and three bottles of white wine, a couple of tins of orange San Pellegrino and some bottled water. 'Should have got some milk,' I mutter and shut the fridge again.

Above the sink and draining board, a window looks out over the back gate of the zoo, toward the muck trailer and the seal and penguin enclosure beyond. The penguins caw as they mill about, the seals up on their rocks to make the most of the late afternoon sun.

Two white-necked gulls go through the muck trailer down below, picking out bits of discarded lettuce.

'Look,' I say to him. 'Seagulls… Probably can't see them,' I add. 'Can't remember when your long-distance vision kicks in. What else we got?'

The cupboards hold plates and cups. Coffee bags. A selection of fruit teas. The draw is full of heavy cutlery and a corkscrew. Three sets of salt and pepper shakers sit next to the microwave. On a shelf above the kitchen cupboards stand ranks of red wine. Provisioned for the hotel guests and all forgotten about when the zoo hit lockdown.

I stand at the sink and turn on the tap so he can see the water. He clutches me as it drums in the sink.

'Sorry dude,' I say. 'It's okay.'

Back in the lounge I switch on the TV, let the background noise roll over us, but he still grunts and gripes, his little legs scrunched up to his belly.

We go out for a walk into the zoo, step into summer early evening light, warm, the air heavy and still. I take him through the gardens, show him plants and flowers. Bees, weighed down with nectar, slowly buzz amongst the stems and gather in a clumsy cloud around a bush of blue lavender heads.

'Let's go see what those sloths are doing,' I say, and we cross from one side of the zoo to the other, keeping away from gorillas, and gradually I feel him relax, go limp and fall asleep.

Back at the flat I wedge him on the garish couch with pillows, tucked up in my hoody. If I was at home with Sparky, she would hold him in case he woke up, or she would get everything prepared, but I'm flying solo. The lounge has a sofa, a large, square coffee table and a big screen TV. An empty mini fridge in one corner and behind the sofa is a glass-topped dining table and six plastic chairs.

I start a chaotic unloading of the once familiar hand-rearing bag. Keep checking he's asleep, microwave my dinner and get back to him as quickly as possible. His body quivers. I see his eyes flicker behind their lids and know he's having a dream. His lips push forward, even in his sleep, and he hoots in fear. Repeated cries.

I put a warm hand on him and grumble.

The hoots continue and his eyes flick open. For a split second he gapes and his face contorts in terror, and I can only assume that the nightmare he was having has become true as he opened his eyes. Taken from his family. He shrieks and wails and I keep him warm and close. Wander the flat again, in and out of the different rooms, wait for the panic to subside and try to distract him with fresh air out on the balcony. It's almost dark. An owl calls somewhere in the gardens.

Eventually he calms down again, and I sit with him against my chest on the couch, wait for him to drift off and dig out my phone to watch the recording of the all-staff meeting.

The announcement that Bristol Zoo will close, and the society will concentrate on their second site: Wildplace. More information to follow in three months' time, lots of enthusiasm about the new gorilla enclosure and miserable finances figures in a confusing graph.

I call Bob.

He picks up on the second ring. 'Alright Al? You watched it?'

I keep my headphones in and slip my phone into my pocket. 'Just seen it, yeah.'

'One hundred and eighty-four years old,' he says. 'Bloody shock, isn't it?'

'It is, mate.'

'Don't see why there can't be a shout out to the public. You seen how much Dublin Zoo got?'

'Dublin?'

'Millions,' Bob says.

'They reckon it'd take millions to fix stuff here, though, didn't they?'

I hear Bob sigh.

'And the problem is the public wouldn't see any difference, it's all behind-the-scenes stuff that needs sorting.'

'You're all for it, then?' he snaps.

'No, I'm not saying that.'

Hasani shudders in his sleep.

'I'm just not surprised by it, that's all.'

Bob sighs again. 'Not surprised by anything these days,' he says. 'Ever since fucking Brexit and what's that bipedal tapir going on about now? Rule of six. What does that mean? Can't go living through another lockdown again, Al. Drive me mad.'

I support Hasani in the crook of my arm. His hands both grasp the vest, but he keeps his toes clenched in cold fists.

'Then you look at the environment,' Bob rambles on. 'What the hell are we doing to it? We're like one of them killer ant species.'

I get up to make Hasani's milk.

'You see them on Attenborough the other night?'

'I missed it.' I push open the fire door into the kitchen.

'Well, you know what I mean,' he says. 'Millions of the little bastards chewing up everything in their path. Leaving a scorched earth trail behind them, destroying everything they come across.'

I'm doing it all one-handed. Hasani is nowhere near strong enough to hold on by himself, so I keep him cradled in my elbow and turn on the tap, slowly so as not to spook him. He remains asleep against me. I half fill the kettle, place it back on its stand and step back to the sink to turn the tap off.

'You'll have some theory on this,' Bob says. 'What do you reckon – humans are the killer ant primate? Billions of us on the planet devouring everything as we go. Actively ensuring our own extinction.' I hear him crack open a can and slurp some beer.

'That's pretty bleak,' I say.

'True though, isn't it? No one gives a fuck about saving the environment. We'll end up as some blip on the graph.'

'What graph?' I switch on the kettle.

'The timeline thing of life on earth.' He belches down the phone. 'You know with the dinosaurs and all that, we'll be some violent blip on there, mammals and mass extinction.'

Getting disposable gloves on one-handed is tricky and my fingers are wet. I wobble Hasani about as I tug them on. His eyes flick open, but he's still dozing.

'Well, this little dude needs us,' I say.

'How is he?'

'Terrified.'

'He's going to be, poor little mite.'

I unscrew the sterilized bottle. 'How are your family doing?'

I hear him swig more beer. 'Yeah, they're alright. What's freaking me out though is if I couldn't come to work. I'd lose it. I mean, what's the point?'

I pop the lid off the milk powder tub.

'I know Hannah wants to see the bright side,' he continues. 'Going on about the new gorilla house and all that, but this place has been here through two world wars.'

Hasani flails an arm and lets out a couple of low hoots.

'Hannah's right, though,' I remind him. 'The closure doesn't necessarily mean we're going to lose our jobs.'

'Keep telling yourself that,' he says and I hear the click of a cigarette lighter. Bob's back on the fags and he'd managed to go five months without.

'We'll know more when they sort out the new species plan,' I say. 'We've got this little one to focus on. Try not to think about the rest of it until we actually know what's what.'

'Easier said than done, Al.'

'I know it is, mate. You got anyone else you can talk to?'

'Trying to get me off the phone?' He breaks into a hacking cough and I wonder if he's holding back tears.

'No. I mean if you're feeling low.'

'Got Monsieur Stella here.' He coughs.

'As long as it's not Comrade Smirnoff?' I say.

'Boris Chesty-cough more like.' He hacks phlegm into his throat.

'Hang in there.'

'Alright, mate,' he says. 'I'll let you get on.'

'We're talking at least two years' time,' I remind him. 'That's what the Higher-ups said in that meeting.'

'I just don't need it at the moment. That's all.'

'Okay. I'll see you tomorrow morning.'

'Night,' he says and ends the call.

I dollop milk powder into the bottle.

'He was cheery,' I say to Hasani. I've heard Bob's rants about the government and Brexit before, of course, his rage about the planet. I feel his gloom sink onto my shoulders and wish Sparky was here. Someone else to talk to and share the load.

I can't believe the zoo is closing.

Life is going to change.

Chapter 17

Single Parent

Lockdown two begins and I am spending more time with Hasani than anybody else in my life. The flat and the zoo are my main residence. I function around his waking timetable. His body has adapted to digest the milk we are offering him, and the feeds get easier, reduced to once every 4 hours now, rather than every 2. If he sleeps, I sleep. The flat is a place of settled routine and provides relief from the continued Covid-19 chaos outside.

The closure of the zoo remains an unspecified amount of time away and I compartmentalize everything. I imagine a multi-drawered sort of filing cabinet in my mind. Everything filed in its particular – and most importantly, closed – drawer. I block out everything other than the day-to-day care of Hasani. Or McCraw, as I christen him.

Wriggles McCraw, to give him his full title. At the beginning, when he was in pain, I would sing, move about the flat trying to ease his colic and would recite: 'There once was a gorilla called Wriggles McCraw. The best of gorillas you ever did saw.' Which I know makes no sense, but I'm on a fucked-up sleep cycle and trying not to think about losing my job.

McCraw is still scrawny, and his fur is dusty and haphazard. He has a ragged hairstyle out of keeping with the other gorillas in the enclosure and is at a stage where, if he were in the group, he would lag hopelessly behind the rest. Despite the mesh between him and the other gorillas, during the day, he remains ever hopeful of kindness, but they generally ignore him. We have to preserve his instinctive urge to belong, but he had never received any compassion from his mum. That's how it feels, and I know my professional zookeeper mind has been torn into ragged tatters, expecting kindness and compassion from an animal is of course attributing them with human emotion, but I can't help it. As I mentioned before, I'd never been a parent of a baby with my kids. Afia had awoken an instinctive side of parenting in me that I hadn't experienced with Fizz or Sam.

Afia had felt like an underdog at the time we hand-reared her, but the trauma McCraw experienced made that pale in comparison. The terror and misery he felt at the forced separation from his group, essential as it was to save his life, means his behaviour is entirely different compared with Afia at the same age. She had been confident and cheery, full of play and vigour. McCraw remains reticent and nervous. I do all I can to make his time with me as happy for him as possible and over the months he begins to improve.

The day has come at last. The announcement about the future of the zoo we have all been waiting for. We are about to watch the all-staff meeting over Zoom that will spell out whether each of us will have a role there or not.

I head to the flat early so I can prepare before the online meeting begins. The preparation is down to an efficient 17 minutes. I leave McCraw with Hannah at Gorillas and zip through the zoo to the flat, kick off my boots and head upstairs. Shower. Bed made with my set of sheets and duvet. McCraw's pillow, just as he likes it. Phone charger plugged in and ready to go. Bottle of water if he wakes up hungry during the night as we're now beginning to cut out his midnight feed. Dinner removed from the fridge and placed in the microwave. Knife and fork in the lounge, out of reach on the coffee table. Toy crate emptied across the floor. Changing stuff set up on the glass-topped table, a production line to enable the swift and effective clean of a shit-smeared baby gorilla arse. Down the stairs, boots on, and I dump the rubbish en route back around to Gorillas. Myself, Hannah and McCraw are to watch the all-staff meeting in the messroom. The rest of the team are scattered in front of different screens across the zoo in socially distanced pods.

The meeting drones on with no real news, until HR talk us through the process of voluntary redundancy.

We wait for the PowerPoint slide to reveal the new staff structure and when it arrives scan the job titles, the different departments and the numbers of keepers required for each section.

'Fuck, Al,' Hannah whispers. 'We're not there. There are no team leader roles at all.'

———

Back at the flat and McCraw and I sit on the couch.

'My job is gone,' I tell him. 'Ceased to exist.'

He clambers off me and eyes the remote on the table next to my knife and fork. I turn on the TV for the background noise and colours as usual. McCraw always has a quiet 20 minutes or so when he first arrives at the flat and enjoys watching Richard Osman's *House of Games*. He leans against the back of the couch, hands behind his head, his feet stuck out in front of him with his toes clasped and stares at the TV, the lights and voices.

An animal department meeting followed the all-staff and the new structure was explained. The team leader role is to be combined with the senior keeper position. The number of keepers across the whole animal team is being halved. Voluntary redundancy is being offered to avoid any compulsory last resorts.

The new species plan revealed none of the small mammal department have made the list of species that will be moving to the new expanded site at Wildplace. There are no plans for a new nocturnal house and my entire team are going to be fighting for jobs. If I want to stay, I will need to go up against Bob and Hannah to remain working with the gorillas.

'Where is Kazakhstan?' Richard asks us all from the TV.

'This is a good round,' I tell McCraw and shelve all thoughts of the announcement away. It's harder than the usual level of compartmentalizing I've been doing recently and threatening to burst out to rain shit all around me, but I still just about manage to ram it all into the metaphorical drawer and keep it shut.

McCraw rests a hand on my leg and keeps his eyes glued to the screen. His other hand reaches for his favourite sofa cushion and he stuffs the corner in his mouth.

'We've got company tonight,' I tell him. 'Seven o'clock.'

'A map of Africa,' Richard informs us. 'But where are these capital cities, please?'

'Right.' I sit forward on the couch.

McCraw looks at me and chews his cushion.

'Antananarivo,' Richard says, and he pronounces it right by the sounds of it.

'Madagascar.' I point to the long, narrow island, about two-thirds of the way up, to the right.

A moped buzzes past outside and McCraw jumps, his hand clutching my leg.

'You're okay, mate.' I rub his back. 'Nothing to worry about.' I gorilla grumble.

'Antananarivo is the capital of Madagascar,' Richard tells us.

'See,' I say. Most of the people on the show have missed Madagascar altogether.

McCraw looks up at me over his chewed cushion.

From his early days of clinging fear, we've built up an expanding daily repertoire of play behaviours together. Slow and reserved at first, a gradual exploration that allowed him to abandon his irrational jumps of fear, the hugs of terror and shit all over the place, a phase that we never experienced at all with Afia. I built up his confidence, step by step, which couldn't help but amplify a deep protective love for him. There is no other way to describe it. I want to help him, make him thrive, and I am spending more of my lockdown life with him than anyone else.

It is working too.

I'm aiming for a good half-hour of charging around at this time of day, the TV forgotten, until he gets hungry for his next feed. I'm not letting the news today push us out of his routine. I slide off the couch to stretch out on the floor, next to his climbing frame.

We've built him a frame of wooden bars and dangling toys, to encourage him to develop and climb.

He stands on all fours to peer down at me from the couch.

'There once was a gorilla called Wriggles McCraw,' I say.

He opens his pink mouth in the beginnings of a play face. I can't properly reciprocate with an encouraging grin of my own as I'm still wearing a mask for fear of Covid-19.

'Kinshasa,' Richard says from the TV. 'Can you find me Kinshasa, please?'

'That's your neck of the woods,' I tell McCraw.

He lowers himself off the couch and reaches an arm briefly for the TV remote.

'Do you want to change channel?' I ask him. 'Thought this was your favourite.'

He drops onto the floor and makes his way behind the coffee table to peer at me and the mini fridge in the corner.

Still here from the hotel set-up, the mini fridge sits in one cramped corner of the lounge and was once an object of deep suspicion. Just the sight of it, lurking behind the armchair, was enough for nervous glances and occasional hoots.

McCraw pauses at the far corner of the coffee table and his scraggy body jerks as he weighs up if fun is on the cards or life is about to deal him another blow. He takes a step toward me, unsure why I'm lying full length on the floor, and keeps hold of the coffee table with one foot for security. He tries a full play face and I grunt and chuckle gorilla laughter.

'Huh huh huh. There once was a gorilla called Wriggles McCraw,' I add.

He trots forward, convinced.

'The best of gorillas you ever did saw.' I present him with the opportunity to slap my forehead in exchange for neck tickles, but we're still working up to that and he creeps up and pats me instead.

I lie on my back and gently rattle one of the metal rings we've attached to his climbing frame. He's currently fixated with the way they move and clink against each other, three steel rings, threaded through a sawn length of broomstick that makes up one of the climbing frame cross-pieces.

McCraw grunts as he clambers over my chest onto the climbing frame and begins to clang the rings together with increasing enthusiasm.

'And the final capital city of this round,' Richard says. 'Can you find me Lusaka?'

I glance at the TV.

He lowers himself to the floor, eyes fixed on the mini fridge again. It's turned off and empty and he now delights in opening the door, seemingly expecting it to have been refilled.

'Lusaka? I know this,' I tell him, but I'm not sure I do.

He makes his way to the fridge door and looks at me over his shoulder as he pulls it open.

'It's empty,' I say.

He pokes his head inside, sniffs and turns around so he can sit down, with his legs dangling out, feet resting on the floor and picks up a teething ring to chew on.

'It's always empty.' I offer him the box of cards, laminated animal facts, some sort of Top Trumps game, also left over from the hotel days. I've got a vague paranoia that he'll somehow scramble out of bed at night, shut himself in the mini fridge and suffocate.

'Come on, you,' I say. 'Out you get.'

He lets me encourage him out of the fridge and turns the box of cards out onto the floor. Scatters them as he scurries toward the leads and plugs under the TV. He's at ease, busy doing what he's doing and has momentarily escaped his nervous cloud.

'Answer Smash,' Richard announces, and I've missed finding out where Lusaka is. 'But first let's take a look at the scoreboard.'

I draw my fingers up and down one of the corrugated plastic bottles which dangle from his climbing frame. The raspy noise piques his interest, and he bundles back toward me again. He's getting hungry, I can tell by his pace.

'Shall we get your bottle ready?'

I kneel up next to him to signify I'm making ready to leave and he grabs for my string vest, rides on my hip as I take him into the kitchen and swing him up onto my back, so he can grip my shoulders. I get a tickly shiver down my back as he shoves his muzzle in my ear.

His bottle is ready to go, all I need to do is warm it up. The feed runs like clockwork. Contented. Body-shaking burps as I wind him, and his eyes begin to roll in his head toward the end of the feed. Milk drunk and bleary. I get my own dinner heated up and leave him asleep on the couch, wedged with the usual cushions and my hoody. Get everything stacked away and set my phone up on the stand Sparky got me. Squish on the couch next to him again, so he never knew I was away, and at seven o'clock, join the Zoom call Hannah has set up for us all to debrief. Time to let the new staff structure drawer spring open again. Talk it out, so at the very least it becomes easier to shove away again in my mind later.

Hannah appears on the screen, sitting on her bed in a big jumper and her beanie hat.

'Hi Al,' she says. 'Is McCraw asleep?'

She wears a giant pair of headphones.

'He's been out for about half an hour.' I pluck my phone off the stand to show her a close-up of McCraw's sleeping face. He mutters and squeezes his eyes. His fingers momentarily clench my vest.

'Bless him,' she says. 'Where's Bob?'

'He finds anything IT problematic.'

Hannah has a picture behind her bed, I can't quite make it out, but it appears to be the stripy back legs of an okapi. A lamp, off screen, throws a speckled pattern up the wall and I can see the fronds of a pot plant next to her bed.

'Weird seeing you up close without a mask on,' I say.

'I'll message Bob,' she says and plucks her phone from somewhere off screen.

'How do you think he's doing?'

'I dread to think.'

McCraw sighs in his sleep and I cover his shoulders with my hoody.

I hear Hannah's phone ping. 'He's just logging on now.'

'At least some of your animals are moving to the new site,' I say, meaning the gorillas.

Hannah picks up a huge mug and blows on it. 'I feel really shit about it,' she replies. 'This is going to spin Bob right out.'

A third screen pops into view and Bob's head appears. He frowns out at us, eyes fixed. Says something, but we can't hear him.

'You're on mute,' Hannah says.

Bob mouths *Fucking thing*.

He's got a bookshelf behind him, stuffed full. Some of the books are upright, others wedged on top of each other in piles.

'... only gave me a lesson on all this bollocks the other day.'

'Hi mate,' I say.

'You've done it.' Hannah's voice chimes like she's training a gorilla.

'Can you hear me?' Bob asks.

Hannah gives him a thumbs up.

I nod. 'We can, mate – how's it going?'

'Took me ages to get home,' he says. 'Queueing up for bloody tests, got us a shed load, though.'

'Legend,' I say.

'I've got some more coming too,' Hannah adds.

'Anyway, never mind that,' Bob says. 'Wait. I can't stand people seeing me.' He fiddles and peers at the bottom of the screen, then sits back in his chair. 'You two don't mind, do you?'

McCraw sits up. Lurches forward and his eyes blink open.

'Hey dude,' I say and rub his belly.

Both eyes slide shut again and he leans back against the cushion, his forehead creased in a frown.

'How's McCraw?' Bob asks and swigs from a can of Thatchers. We can still see him.

'Crashed out, think he's having a dream.'

I manoeuvre my phone, show them both his wrinkled face.

'Probably having a shit,' Bob says. 'Here, have you been for a piss with him on board?'

'Not tonight,' I reply.

Hannah sips from her giant mug.

'Little bastard grabbed me old fella last night, swung him about as I was pissing.'

'What?' Hannah shrieks.

Bob laughs. 'I pissed all over the floor.'

Hannah grimaces. 'Gross.'

'Yeah, sorry Al. I mean I cleaned it up like, when he was asleep later, but he's well into the bathroom at the moment.'

'Why don't you put him down in there?' Hannah asks.

'Can't,' Bob protests. 'He got himself stuck behind the sink pipe and he took a bite out of the toilet roll.'

'I'll look forward to that later,' I say.

'Used to be terrified, didn't he? When you flushed the toilet, but now he's got to get an eyeful and he was flailing about all over the place.' He takes another slug of cider. 'You telling me he doesn't try that with you two?'

'Bob.' Hannah sighs. 'I don't piss standing up.'

He laughs. 'Oh right, yeah. Don't you?'

'Why don't you sit down too…' Hannah shakes her head. 'I can't believe we're talking about this.'

Bob's trying to put off talking about the all-staff meeting for as long as he can, even though that's why Hannah organized the call.

'I'm getting another drink,' he says. 'If I don't answer, I'll be back in a minute.' He stands up and disappears off screen.

'That's my copy of *Marsupials and Monotremes*,' Hannah says. 'Second shelf up.'

I squint. 'Is it?'

'I wondered where that went. It was in the messroom years ago, and is that *Modern Zoo Design*? Top shelf?'

'I can't see it.'

'Underneath *Lemur News*.'

'I'm back,' Bob says and slumps back down in his chair. 'Come on then, Al. How you feeling, mate?'

I haven't put together how I'm feeling. Haven't had time.

'All my animals are going,' I say.

'Fucking disgrace,' Bob growls.

'I don't know, is it?'

'Course it bloody is.' Bob takes a savage gulp from his can.

'Nocturnal houses need a shake up,' I say. 'It's hard to argue that an animal should spend its whole life indoors. Hard to justify that on welfare grounds.'

'Best nocturnal house in Europe,' he responds.

'People aren't interested,' I say.

'Some are,' Bob argues. 'Right Hannah?'

She nods.

'Don't get me wrong,' I say. 'I couldn't be prouder of Twilight and the team. The studies they're all doing, UV stuff, the lion training Scott's doing... and I know some of the public like it, you know the interested ones, but the bulk of visitors spend about two minutes in there. Say they can't see any animals.'

'Stupid arseholes,' Bob mutters.

'I'm not surprised nocturnal species haven't been included in the collection plan for the new zoo is what I'm saying. I'm gutted, obviously, particularly about the aye-aye.'

'Good luck moving everything,' Bob says. 'That floppy-haired bipedal tapir has fucked all that right up for us.' He waves his arms, ready to launch into his Boris Johnson rant. A clatter of what sounds like empty cider cans comes from off screen. 'Balls,' Bob adds.

He's right. Brexit has turned the task of emptying the zoo into a logistical nightmare. Moving animals to Europe was complicated enough at the best of times and now we have the post-Brexit rules, awaiting adaptation in some monstrous pile of pending legislation, different for each country and on top of that a global pandemic.

Bob's usual ready smile has disappeared, it has paid infrequent visits to his face, even when we see him without a mask on. Ever since the announced closure of the zoo back in the summer, his whole persona has tipped over into bitterness and he has been awaiting this day with dread, just like the rest of us.

'How many European breeding programs has he fucked up?' Bob leans off screen to retrieve his foaming can.

Thoughts on all the future animal moves hasn't entered my mind yet, all consumed in McCraw and his needs. He's my buffer against the pandemic, nights away from my family and now potential redundancy.

'It's going to be a nightmare,' Hannah agrees.

'Should never have bloody happened,' Bob shouts. He's shit-faced, I realize. We wait for Bob to work through his rage. 'Bloody bus with the NHS lies. And people actually fucking believed it.'

'How are you finding it?' I ask. 'Being at home alone?' I can't listen to his tirade about the ruined environment tonight too.

'Prefer it when I'm there with him,' Bob says and slumps in his chair. 'Even if he does try to grab your cock.'

'Bob,' Hannah chides.

'It's just shit,' he says.

'Are you seeing anyone?' Hannah asks him. 'Your family?'

'Yeah. Me niece gave me a lesson on this fucking Zoom bollocks the other day, through the cat flap. Dropped them off some shopping.'

'What about you, Al?' Hannah asks. 'How's the family coping with you away three nights a week?'

'Okay. Sparky played such a big role with Afia, it's a shame she can't meet McCraw. She gets it. We have to limit the

Covid risk, but she really helped out last time. How single parents do this I don't know.'

'Tells you everything you need to know about this country,' Bob says. 'Why does everyone hate single mums so much?'

Hannah and I both nod.

'It's so different, isn't it?' I talk over him by mistake. 'Doing this alone.'

'That's what I'm saying.' Bob belches. 'We've got a society that abandons single parents, at the point where they need all the social support they can get, but everyone isolates them and ignores them instead. Makes them out to be public enemy number one and now another fucking lockdown.'

'There'll be a vaccine soon,' Hannah states. 'This won't go on forever.'

'I'm so fucking lonely,' Bob says. 'Get home, shut the door and that's it.' He crunches the empty cider can in his hand. 'Nothing, no one to talk to. TV driving me bloody mad...' He tosses the can off screen and we hear the tinny rattle as it lands with some others. 'I'm getting another drink.'

McCraw kicks me with his little feet. I feel his warm chest, slide my hoody half off him.

'Are you doing okay?' I ask Hannah.

She smiles. 'It's a shock. The restructure. I feel terrible for you, Al, all the animals you manage.'

We hear Bob crash about off screen.

'I'm not thinking about it yet,' I say. 'It's not like it's happening next week. We've got to get this one back with his mum first.'

Bob returns, lowers himself in slow motion into his chair. He inhales on a cigarette, the burning end gleaming on the screen.

'At least Kala's still interested in him,' Hannah says, and we repeat the conversation around mixing plans that we have at work every day but can't stop ourselves going on about it again.

Bob stays silent, distracted, and his eyes roam around his room. He pokes a finger into one of his ears.

The underside of Hannah's face illuminates. She's picked up her phone off the bed. Peers down at it.

Bob finishes his cigarette and gently lifts a shot glass to his lips. We see him wince as he clinks it against the ashtray.

My phone buzzes. A message from Hannah appears on my screen.

'He thinks his camera's off, right?'

Bob downs the shot and gently places the glass off screen.

My phone buzzes again.

'Is he pretending not to be there?'

'I'm back,' Bob states, 'just had to take the rubbish out. What did I miss?'

'Nothing,' I say. 'Anyway, back to your original question. I don't really know how I'm feeling about it all. My role has gone from the staff structure. I can't get past it.'

Bob sighs, shakes his head. 'One of us three will be out of a job for sure. Fucking not bloody fair.'

'We still don't know that,' Hannah says.

'Course we do.' Bob sniffs. 'Whole working life at this place and now it's closing. Alright for you two, with your degrees. What am I going to do?'

'People might choose to go between now and then,' Hannah says. 'Take voluntary redundancy, find jobs somewhere else and leave. It's still ages away yet.'

Bob's face opens in a silent howl. He shoves his fist in his mouth and deliberately slaps his cheek.

'Whoa, hey Bob,' Hannah cries.

He freezes.

'All they did today was announce the new staff structure,' she says. 'We're still talking over a year's time.'

Bob stares out of the screen, his face drawn, tears in the stubble on his cheeks. 'Can you see me?'

'We can, mate,' I confirm.

'Fucking thing.' He slumps forward. 'Thought me camera was off. I can't see me my end.'

'Are you okay?' Hannah asks.

'Clearly fucking not, if you can see me.' He breaks down into big blubs.

'Come on, mate.' I try to console him. 'It's the shock, but like Hannah says, we've got loads of time to think about it.'

'The zoo will be gone, though,' he wails.

'Wish I could hug you,' Hannah says.

Bob holds his head in his hands. His shoulders shake and shudder.

'Listen.' Hannah leans closer to the screen, her face scored with lines of concern. She saves this look for animals under anaesthetic, like when Kera was at death's door all those years ago. 'The gorillas need you.'

'Hannah's right, mate,' I add and switch the camera on my phone around, show McCraw's sleeping face, adrift and carefree.

'We need you, Bob,' Hannah continues, and she's teared up too, her voice catching in her throat.

Bob blows out a big lungful of air. 'Everything is just so fucking bleak,' he says. 'Never any good news, about anything. The environment. All the European breeding programs and those privileged, blubbery twats have fucked all that up.' He slumps again. The rage bubble pops before he gets going. 'I can't see no one again and it's all changing.' He nods. 'Change. I bloody hate it.' He pauses. 'Wait, did you see me smoking?'

'Yeah,' I say.

'And poking your ear,' Hannah adds.

'I didn't know you could see me,' he protests. 'Lucky I got clothes on.' He doesn't deliver this line like he usually would, can't find the enthusiasm.

'It'll work out, mate,' I say and feel a wave of his fear slosh over me, a ripple of Bob's gloom. We're going to have to go up against each other for jobs if we want to stay. The gorillas are moving to the new zoo, but the majority of the animals we work with aren't. Half of us won't be going either.

'Got to get you back in the group first,' I say to McCraw, who is still fast asleep slumped against my leg on the sofa.

'I'm really sorry,' Hannah says, sighing, 'but I need to go.'

'Okay,' I say.

'Hang in there, Bob,' Hannah adds. 'Call me if you need to, and I want my copy of *Marsupials and Monotremes* back.'

'What you on about?'

'Night.' Hannah waves.

McCraw sits up as I end the call. Reaches out his arms for me and I pull him up onto my chest. I need a piss of course. Best to go now, while he's still sleepy.

His little body is warm. He curls his toes around my belt and grips my vest with one hand, trails his fingers to tug the door as we head out of the lounge.

There's a big round mirror in the bathroom above the sink. The light flickers and the extractor fan rattles into life. I could do with cleaning my teeth, but first things first.

He catches sight of the gorilla in the mirror and presumably the other version of me that kicks around this flat. He scurries around me with clutching toes and fingers so he can pat the gorilla's hand in the mirror.

As soon as I start pissing he forgets his image and tries to slide down my back. Four determined, strong limbs, all in action as once, as he tries to clamber around my waist, craning his neck to catch sight of the stream of urine. It's impossible not to piss all over the floor and the underside of the seat. Something I've prided myself on not doing for years.

I grab him with one arm and he immediately does a deadweight drop, clinging to my forearm with both hands, while he tries to snatch and grab with his feet.

'Pack it in.' I throw him up onto my back and give up on my piss. 'Going to have to clean the floor now,' I mutter and see myself in the mirror.

All my animals will be going to other collections. My entire small mammal team are at risk of redundancy.

Everything we specialize in will be gone.

McCraw looks over my shoulder.

'Except you,' I say.

Chapter 18

Covid Christmas Day – at the Zoo

I creep out of the house with an overnight bag on my shoulder and a carrier bag with McCraw's present, along with a full portion of Christmas dinner and gravy sealed up in a plastic tub with clingfilm. I'd cooked Christmas dinner yesterday for Sparky and Fizz, celebrating a day early as I'm off to work. Fizz moved back in with us to ride out the second lockdown and Sam remained over in Belfast. I've set an alarm on my phone for our Christmas Day Zoom later.

The roads are deserted, apart from two lonely runners pounding the dark pavements, presumably as usual. I park in the near-empty carpark next to Twilight Scott's moped. It's 6.30 a.m.

Hannah waits in the messroom to handover McCraw. She's got a new face mask, a reindeer snout with a red nose.

'Happy Christmas,' she says. 'I've put the coffee on.'

One day a year the zoo runs on half numbers. A double set of diets were made for all the animals yesterday. Cleaning is kept to a bare minimum, the animals are checked and medicated if required, given fresh food and water and left to their own devices for the day. The cleaning is all picked up on Boxing Day. If all goes to plan, most keepers get off site by 9 a.m. to disperse and have their Christmas Days back at home.

'Merry Christmas,' I say and McCraw waves his arms at me.

'He had some water overnight,' Hannah says. 'But apart from that he slept through.'

'Hello you.' I hold out my arms and he climbs off Hannah.

I've worked more than my fair share of Christmas Days over the years. As I do the rota it always felt horribly unfair to be off while my colleagues were in and the zoo-wide rule was you always got the following day off to make up for it, so I never really minded doing it. This year it was all different of course because of McCraw.

I work flat out until four and I realize as I bring McCraw back to the flat that I've missed any chance to have a shower.

I set a reminder alarm to check the Twilight World boiler and take McCraw into the lounge.

I've got our family Zoom later, with Sparky and the kids. Nothing for a couple of hours though and a Tarzan movie about to start on the TV, rather than McCraw's usual *House of Games*.

I lie on the couch. McCraw sits on my lap and we watch the bizarre-shaped gorillas walk into a tree house and take a baby Tarzan out of his crib.

McCraw absentmindedly chews the corner of his cushion.

'Look at the way they move,' I say to him. 'Bit stilted, aren't they?'

He stares at the TV, the gorillas swinging through the trees on long ropes, Tarzan, now a kid, swooping along right there with them. McCraw takes the cushion out of his mouth.

'Their arms are held too straight when they walk,' I say. 'Bit like cartoon ponies.'

His mouth opens in a yawn and he sways, drops his cushion on the floor and pulls his way across my chest, to wedge his head under my chin, just like Afia used to do, the soft gorilla fur tucked against my neck. We last about 2 minutes before we're both asleep.

I wake up to him laughing. He's rolled off my neck into the crook of my arm, his body shaking with hysterical chuckles, and I realize he's still asleep. Dreaming of something hilarious. I get a rush of euphoric joy, tinged with a bittersweet edge as I realize how infrequent that feeling is anymore. I revel in it and I know when I pass it on to everyone tonight on Zoom, they'll feel good hearing it, an additional boost to any Christmas cheer.

He continues to chuckle and his fingers and toes twitch.

I think back to that first night with McCraw, when he was the skeletal thin Hasani and his waking nightmare, screaming and shrieking in terror, and now here he is four months later, cackling and smirking. The chuckles subside and come back again into full hysterics. His eyelids flutter and his lips twitch. He opens his eyes and looks at me with a cheery play face, reaches out and pats my chest.

The Tarzan movie is now a fight on a kind of cowboy Wild West train.

'Right you,' I say. 'Let's check the boiler in Twilight World.' I zip him up inside my hoody and head out into the cold zoo. 'You can open your present when we get back.'

The heating is on. Nothing tripped.

I take him back to the flat, stick my phone on the stand and join the Christmas Day Zoom.

Sparky and Fizz have glasses of Baileys on the go and Sam, over on a screen in Belfast, wears a hat with earmuffs.

They cheer as I appear. I'm 5 minutes late and a chorus of 'Merry Christmas' greets us.

McCraw leans forward to look into the screen.

Sam laughs. 'Hello, Mác Claws, no wait, what's his name again?'

'Hasani,' I say, 'but I call him Wriggles McCraw.'

'How's today been?' Fizz asks.

McCraw yawns and pulls himself back onto my chest.

'Went a bit crazy,' I tell them all. 'Bat emergency and I had to call the vets in. The boiler kept going down in Twilight World, but I had Scott to give me a hand and everything's done now.' I realize I haven't eaten yet or offered McCraw any food.

'Is he going to sleep?' Sam hasn't seen McCraw yet, I realize. We normally Zoom when I'm at home.

'He's been riding around on my back all day, up to the vets with the bat, came to feed the seals and loved it up at Twilight World, looking at all the off-show rodents.'

'What's an off-show rodent?' Sam asks.

'In the Twilight World keeper-corridor there's groups of rodents, back-up populations. He was obsessed with them. If it wasn't so cold I could have taken him on an afternoon tour of the whole zoo.'

'Aren't you worried he'll run off?' Fizz says.

'He's being a proper gorilla,' I say. 'If he's somewhere unfamiliar he sticks on my back, just like he would with his mum.'

'This is the fourth little kid you've brought up, Al,' Sam says. 'Right Fizz?'

'You're right.' Fizz laughs. 'You and me are numbers one and two, Afia three and now McCraw, number four.'

'I'm going to microwave my dinner,' I say. 'Keep talking, I can still hear you.'

McCraw looks around, fingers and toes clutching my string vest.

Everything is already prepared. The microwave hums and the orange light attracts McCraw's attention. I hear my family in the other room and for the briefest second think they're all actually there, chatting and laughing as I get food ready in the kitchen.

I collect a knife and fork and McCraw's dinner and take it back into the lounge.

'What are you having?' Sam asks.

'Christmas dinner,' I say, 'leftovers from yesterday.'

'We had curry,' Sparky says. 'It was nice, wasn't it Fizz?'

Fizz nods. 'It was lush. What's McCraw having?'

'A raw version,' I say. 'Without any meat or gravy.'

I take my phone off the stand and flip the camera round to show them McCraw's tea, ready and waiting, a takeaway carton of chopped carrots, a steamed parsnip, butternut squash, a bunch of basil, and half an iceberg lettuce.

'He doesn't eat all that, does he?' Fizz asks.

McCraw's arm darts into view as he makes a grab for a carrot and shoves it into his mouth.

Sam laughs. 'Remember Afia eating dinner, chucking it on the floor and smearing it everywhere.' Sam had shared a meal with Afia on a weekend he'd been back from university, the only time they'd really met.

The microwave pings and I stick my phone back on its stand.

McCraw has his feet on my belt and clutches my string vest with one hand. He grunts as I stand up and head back to the kitchen.

'What's with the hat?' I hear Fizz ask Sam.

McCraw drops his carrot to make a grab for the Paignton Zoo fridge magnet.

'It is so cold here,' Sam says from the lounge. 'It was a Christmas present.'

I bring in my own takeaway carton of roast veg, chestnut stuffing, crispy bacon and turkey.

'Why is yours in a carton?' Sparky asks. 'That looks so depressing. There must be plates there?'

'It's handy to have a lid.' I rest my food at one end of the coffee table. 'Just in case.'

Sam laughs. 'Mum, do you remember when Afia grabbed your corn on the cob?'

Sparky laughs too, but I can see Fizz still remembers Afia chasing her from the lounge, screaming and showing her teeth. 'I know.' Sparky continues to chuckle. 'Help yourself why don't you.'

'How's Afia doing?' Sam asks.

'She's jealous of this one,' I tell them. 'Doesn't like seeing him have his milk, or her milk as she sees it.'

McCraw's tiny black fingers squish his parsnip and he raises it deftly to his mouth. The days of poking himself in the face with his food are over and he casts a surreptitious look at my Christmas dinner.

'I thought they don't remember being hand-reared?' Fizz heads off screen toward the kitchen.

'You have to continue bottle feeding for three years,' I say. 'Afia definitely remembers her bottle.'

'I was jealous as shit over Afia,' Sam says. 'Weren't you, Fizz?'

'She was so cute.' Fizz reappears with the bottle of Baileys. 'And terrifying. I know you explained why she chased me that time, Al, but I literally thought I was going to shit myself.'

'That was no good,' Sparky says.

'It was so weird seeing you both with her,' Sam adds. 'Not on a serious level, but just odd.'

'You should've tried living here.' Fizz laughs. 'All the cooing and praise.'

'You got usurped,' Sam says.

'By a gorilla.' Fizz nods. 'I was the cute one in the family. The youngest, that was my role, and Afia turned up and no way anyone could compete with how cute she was.'

'Sorry Fizz,' Sparky says.

'Yeah.' I jump in. 'Sorry.'

'I wanted her to love me,' Fizz says. 'So I still had a role in the family, could look after her.'

'Touni took on that role when she first arrived,' I say. 'Buddied up with Kukena to find her place in the group.'

My family all look out of the screen at me. None of them are quite sure who Touni is, or Kukena.

'How long before McCraw has to go back in, Al?' Sam's tone is concerned.

'Not long,' I say. 'He's nearly ready.'

'Will his mum take him back?' Sam asks.

'We hope so.'

'She's a bit of an idiot,' Sparky says. 'Her mum. Inexperienced.'

'If it doesn't work out,' I say, 'it would mean either hand-rearing him until he's sturdy enough to go in with the others, without any back-up, or sending him off to a different zoo and a surrogate mum gorilla.'

McCraw wipes his hands on the couch and reaches for his lettuce.

'Keep him with you,' Sam says, because he knows what a crazy experience I'm having and is pre-empting what he thinks I'd want. 'You know what you're doing.'

'It's tricky,' I say. 'It's all about the rate of his development, alongside being kept on his own for long periods of time.'

Sam nods and the earflaps on his hat shake. 'None of us want to be on our own.'

'We're social primates,' I say, 'feel safer in groups.'

McCraw eyes my food on the edge of the table.

'I wish we were all together right now,' Sparky says.

Fizz lays a hand on her shoulder and reaches off screen for the Baileys again.

I take advantage of McCraw pulling apart his lettuce to shove my dinner down and with a round of 'I love you' and 'Merry Christmas', we end the call.

I get a message from Sparky saying to call again once I've checked the Twilight World boiler one final time.

I clear up and McCraw has his bottle.

We watch TV until it's time to check the boiler. I zip McCraw against my chest inside my hoody. His head and one arm poke out and his nostrils twitch in the night air. Our breath comes out in clouds and the grass stems crunch with frost as I cross the main lawn to Twilight World. In the keeper-corridor he cranes his neck as we pass the glass tank of spiny mice, eager to see them again, darting along branches and scurrying in and out of nest boxes. I know the boiler is running fine by the temperature in the corridor but check it once more to give myself a chance at a decent sleep.

He fidgets and slaps the light switches as I turn them off and we head back out into the dark zoo, our final Christmas Day tasks complete. He's exhausted and as soon as we get back on the couch, he slumps across me fast asleep.

I call Sparky.

'All done?' she asks.

'Yep,' I say. 'Has Fizz gone to bed?'

'I think so.'

'I'm thinking of taking voluntary redundancy,' I say.

'Where did that come from?'

'I was chatting to Twilight Scott earlier,' I tell her. 'When we were feeding the seals. He thinks half the team will take it.'

'Voluntary redundancy?' I can hear the shock in her tone.

'It won't be for another whole year yet,' I say.

'You're not chopping off your nose to spite your face, are you?'

Sparky and I have been together for the best part of 20 years and she knows me inside out.

'Maybe a bit,' I admit. 'I know it's not meant as a kick in the face, but it's all my animals that are leaving, my team. The species we've specialized in for years. I know it's the reality of the new collection plan—'

'You'd get a job at the new site if you wanted it,' Sparky butts in. 'They rely on you for everything,' she adds with a slight edge of bitterness. 'You're there now. On Christmas Day.'

'Yeah, but that's, you know. One of us has to be.' I check McCraw isn't too hot. He shuffles as I lay my hand on his back.

'That's why you feel it's a kick in the face, though,' Sparky continues. 'You do a lot for that place, you can't help feeling under appreciated.'

'I think it might be time for a change,' I say. 'When the world gets back to normal.'

'If it does.'

'We've been vaccinated. That's going to have a massive impact.'

'What would you do?' Sparky asks. 'If you take voluntary redundancy?'

'Something more hands-on conservation wise,' I say and hear a naive edge in my tone, having no real idea what I'm talking about. 'Surely there must be some understanding now, globally, that cutting emissions has a really positive impact on the environment.'

'You think so?'

'I read something about bird song getting more intricate,' I say.

'You what now?'

'Because there's less background noise to drown it out, apparently it should mean that wild bird song, to each other, you know, when they're attracting mates, will be able to expand, become a more nuanced language, for want of a better word.'

'Does Bob still go on about all that killer ant stuff?' I hear Sparky yawn down the phone. 'That we're all doomed.'

'Yeah, give him half a chance. That's another reason to take voluntary redundancy too,' I say. 'Freeing up a space for someone else.'

'Meaning you don't have to go up against them at interview?' she says.

'Maybe. I've done two columns, in my head. Tried to write them down, but McCraw kept grabbing the pen. For and against.'

'I've had better Christmases.' She chuckles. 'Talking about work?'

'Do you want to hear them?'

'Sure,' she says.

'Pay,' I say.

'Terrible,' she replies. 'Nowhere near the national average wage.'

'Hours.'

'Great, if you want to miss parties and weekends away,' she says. 'Time with your family, weddings, holidays longer than twelve days in a row.' Sparky gets particularly irritated by the way our rota works.

'The hours are in the pros column,' I say.

'What?'

'Most zoos work a system of two weekends out of every three.'

'Your industry is fucked.' She yawns again down the phone. 'Ruthless. They take advantage because you're all bat-shit crazy for your animals.'

'Maybe we should look at moving again. I'm going to be working somewhere else, that's for sure. Not here in Bristol, I mean, so maybe we should look at what we want?'

'Bloody hell,' she says.

'Sorry, I'm just rambling, following a thought process you know. This part of my life is coming to an end. I could stay probably and run a section, but it wouldn't be gorillas and none of the species I find really interesting are going to be there.'

'You always say the animals are as interesting as you make them. Always something new to learn.'

'Yeah,' I agree. 'That is true, but I've specialized, should be looking all over the place, if I'm going to stay in zoos.'

'Could you leave the gorillas?'

I look at McCraw. He's in his deep sleep phase, gone. 'He'll be in the group by then,' I say.

'You'll still have to leave him, and what about Afia and Kera? You love Kera. And she loves you. You're a key part of her life, don't forget. Who else has she got? Jock still doesn't like her, does he?'

'Not really.'

'Afia doesn't even like her, does she?'

'No. She was too rough with her that time and she's had nothing much to do with her since.'

'That's what you've got to think about,' Sparky says. 'You don't want to mess yourself up leaving all of them behind, do you?'

'No.'

'That's why I asked if you were cutting your nose off to spite your face.'

'Yeah.'

'Merry Christmas,' she says.

Chapter 19

Mixing with Kala

McCraw is ready. We're ready. We've done this before, got Afia into the group with Romina, and this time we are introducing McCraw to his actual mother. There's an undeniable maternal bond there and it's time to get him back with his mum. The longer we keep him out of the group, the harder it will be for him to assimilate and become a gorilla.

The first mixes are short, as we still have to fit in the daily gorilla husbandry routine, the cleaning and feeding. For the first week McCraw is game to give it a try each day, but he becomes increasingly distressed when we leave him with his mum. If we stay in close proximity to him, he wobbles our way hooting, so we step out altogether and watch each mix on the monitor in the messroom. Daily, I begin to engage in a monumental battle, taking cover behind a shield of zookeeper professionalism, to hold the gut-wrenching panic and fear for McCraw's safety at bay. Dry tongue, sweat that runs in cold streams and constant clock-watching, longing for the moment we can end the mix on a positive interaction and get McCraw out again.

It becomes clear that Kala is going to be problematic. Whereas Romina, back when we started to mix her with Afia, was happy to spend time away from the rest of the group, Kala is not. Romina was 36 years old when she became a surrogate mum to Afia. Kala is only eight and is still half a child herself. She doesn't want to be separated from everyone else, and most importantly Jock. Her place in the hierarchy is shaky, close to the bottom, and she needs to improve her position, pays constant attention to where everyone else is and sits close to Jock, blocking Trim Touni and Kera when she can and dissolving back into her childhood to play with Afia or Ayana if they let her. She is missing out on all of that by being separated with her son.

Touni, as head female, is getting all the company from Jock she requires and even Kera is comfortable sharing a den with everyone else. Afia and Ayana are playing together and Kala can't bear it. She adopts her manic pace almost immediately and whereas she seizes McCraw as soon as she can and he is happy to be onboard, Kala begins her repetitive forward rolls and squishes him with her entire weight against the floor each time. The longer Kala is separated from the others, the more erratic her behaviour becomes, and the mixes over that first week become increasingly brief.

The emotional battering gets worse. Abandoning McCraw each day makes me feel sick. Physically sick. I know it has to be done, that we are aiming for a specific goal, McCraw reintroduced to the group, but the dread settles in when I wake up, knowing this unsuspecting happy gorilla is about to be left at the mercy of an incompetent, inexperienced animal with an anxiety-inducing fear-of-missing-out syndrome. It goes against the grain of everything I feel for him. I am aware my emotional bond is influenced by the chaos of life in general, the ever-present fear of passing Covid on to my animals and the uncertainty of continued lockdown and the closure of the zoo. I am aware I am carrying all that, but placing this trusting little gorilla in danger each day begins to feel like utter betrayal on my part. I have to accept a mantle of deadened compassion.

The closest I came to inflicting misery to this extent on my own kids was taking Sam off to camp one summer. It was advertised as kayaks and rock climbing, but as we approached, I sensed Sam didn't want to go, but he trusted in me enough that despite the anxiety and terrible weather, it would all be exciting and full of adventure. Years later Sparky and I learned how awful it had been, with Sam subjected to bullying and shouting instructors. The crushing guilt I felt then floods through me now.

Just like Sam, the relationship I have with McCraw is based on trust. I got him to overcome his initial terror at being taken from his gorilla family, provide constant security, and despite being different species, we understand each other. His bafflement as I turn my back on him each morning and actively leave him behind, ignoring his hoots and cries of panic, to disappear from his view altogether, that weighs heavy, and as each day rolls on I try to pull the shredded tatters of a professional zookeeper back about myself to avoid becoming a gibbering mess.

Touni is added to the mix in the hope that company will help lower Kala's stress levels. Touni has given birth to her own son Juni and is showing Kala how it is done. Things stabilize initially, but as the two of them are now both separated from Jock, tensions rise. We make the call to add Jock to the next mix, to see if his presence will calm both Touni and Kala down.

I carry McCraw along the upstairs keeper-corridor. He cranes his head to see who's where and I can feel the rising tension in his hands as his fingers grip my arms.

'It's alright,' I say and give him a squeeze.

His clutched toes tug the string vest taut, at the dawning realization that he could be about to get abandoned. Yet again.

'You'll be okay,' I whisper, which is far from certain.

I sigh, try to block it all out with the here and now. A functioning gorilla. That's the bottom line.

The red fire hose has been coiled up all wrong, back on itself, and if Kera had access up here she'd be trying to fish it through with a bit of stick. But Kera, Afia and Ayana are shut away in den two, picking up seeds and Old World monkey pellets, scattered liberally to keep them busy.

I unlock the end feeding den and swing open the heavy door. The Sharpie line on the floor, denoting how far Kala can reach her arm through the gap we're about to open, is scuffed and faded. I step in and kneel on the far side of the line, like I did with Romina and Afia all those years back. Let McCraw unhook his toes from my string vest and place his feet on the concrete floor.

Kala comes for us straight away. She strides the row of feeding dens, crackling with the desperate maternal need for

McCraw and the suppressed fear at our new daily habit of allowing an arm-sized gap between us. No mesh. Just the slide from one feeding den into another, open the correct width for McCraw to get through, but not wide enough to let an adult gorilla in on us.

I double-check the Sharpie line, gorilla grunt and grumble reassurances as Kala approaches. Her long face makes her eyes appear slightly too close together and gives her a solemn look, but her hairdo, the crest of fur off the back of her head, reminds me of Romina, neat and fanned.

'Channel your inner Romina,' I say and gorilla grumble some more.

She leans her shoulder to the gap at the slide and reaches through a long, thick, hairy arm. Her fingers are well over the line and I get both a surge of fear and the instant crisp rush of adrenalin, ready to jump away.

McCraw reaches forward to pat the top of her big grey hand and she grabs his wrist. Pulls him through the slide all in one smooth movement and slings him on her belly. His little feet scrabble for purchase on her fur as she swiftly about-turns on her bandy legs and takes him straight out onto the upper level in den one.

I step out and quietly pull the keeper-door shut and hook the padlocks through but don't clip them closed. Kala grabbing him like that has spared me the moment of abandonment we've been inflicting on him all week. I join Bob and Hannah in the messroom. My mask is wet and smells sweet. I try and ignore it.

Hannah has her red panda-print face mask on, whiskers and nose that looks more like a ferret or mongoose. She sits in her usual corner of the messroom, eyes glued to the big wall-mounted monitor.

Bob is at the desk. He has rediscovered one of his original masks from the early days of Covid that sports a plastic nozzle on one side. He has control of the camera feeds up on the computer and tugs the mouse out from under some piled uniform. A fork bounces onto the floor beneath his chair and yesterday's plate clanks next to him. 'Fuck's sake,' he hisses and plucks a pale orange spaghetti hoop off the scaley skin of his elbow. 'Who left that there?'

'Don't,' Hannah pleads from the corner. 'It's so disgusting.'

Bob lifts his facemask and sucks the spaghetti hoop into his mouth. 'She don't know where to go look.' He nods at his screen.

'Romina is who we need in there,' Hannah states. 'She knew what she was doing.'

I sit between them both to maintain social distancing and look up at the main monitor on the wall. Kala paces the length of the upper level. Jock lies on his front, propped up on his elbows, beneath the UV lamps and glowers at Kala as she approaches, McCraw clinging to her belly, just visible on the screen.

The monitors all flicker at once and the messroom lights dip.

'That's all we bloody need.' Bob shuffles forward in his chair and reaches across the tangled radio chargers for the clipboard. 'If the cameras trip out now we're fucked.' He sticks his elbow on the plate again reaching for a pen.

'Good she picked him up straight away,' I say.

'Did he hoot?' Hannah asks.

'Yeah, but not as much.'

Touni sits up on top of the hide, the glass ceiling which covers the empty public area. No public have been in since lockdown two began, but we'd have locked them out of the house anyway so as not to risk someone filming the gorillas if things went horribly wrong. No one wants to see social media footage of a baby gorilla bludgeoned into a broken-boned mass or torn quite literally limb from limb.

Jock leans forward on his huge forearms, a sweep of seeds gathered in front of him, and keeps his beady eyes on Kala.

'Do you reckon Jock can see him?' Bob asks.

Hannah's chair squeaks in the corner. 'I don't think so.' She pulls off one of her wellies and lies it on top of the radiator. 'He's incapable of recognizing babies as gorillas at this point, too small for him to fathom they're part of the group.'

'Fuck me.' Bob sniffs. 'Is that your feet? Thought Jock's arse smelt bad this morning, but that's something else.'

'I've had pissing wet feet all day,' she snaps.

'I can smell them all the way over here. Through me mask.'

'My boots have got holes in.'

'Get some new ones.'

'These are my new ones.'

'Can you smell that, Al?' Bob says.

I wish they'd both shut the fuck up.

Kala lies on her back.

Jock ignores her.

Hannah props her second wellie in line with the first on top of the radiator. 'Coffee?'

'Please,' I say.

'Better chuck half a bag of sugar in mine an' all,' Bob says. 'Could be a long night.'

Hannah yawns. 'It's already been a long day.'

I glance at my phone; it's almost four o'clock.

'So far so good, though.' Bob keeps his eyes on the computer screen. 'Calmer than yesterday with Jock in there, look.'

'Hope Kera doesn't terrorize the other two.' Hannah pads around the corner to the kettle, her mismatched wellie socks leaving damp footprints on the dusty vinyl floor.

Kala rolls back to her feet again and Touni glances her way but remains focused on picking up seeds, one at a time, on the end of her licked finger. Her little son Juni, only a month or so younger than McCraw, begs at her face, lips trying to reach hers. She ignores him, each dab of her finger swift and precise.

I christened Touni's new baby Perfect Juni, after the character in the *Horrid Henry* books. *Perfect Peter.* On long drives to Europe when the kids were little, camping in the summer holidays, *Horrid Henry* would often accompany us on a CD in the car. Collectively as a family we would all boo *Perfect Peter.*

Life was so sweet for Perfect Juni. Touni wasn't the most doting of mothers, but she knew what she was doing, and he had his big sister Ayana to look out for him too and play with. His fur was neat and groomed, sleek and well turned out, whereas McCraw, in comparison, was always dusty, his coat a bit clumped, no matter how much we tried to groom him.

'There's another cockroach in the percolator,' Hannah snaps. 'Who made the last batch?'

I try to fade them both out and keep my eyes on the big screen.

'Thought me last coffee tasted oily,' Bob says.

'You're so vile,' she hisses.

Kala takes McCraw downstairs, slung on her belly. Her bandy legs are squeezed together as she goes down the steep set of steps and we lose sight of her on the main monitor.

I stop myself snapping at Bob to hurry up and split the screens and hear him click away with the mouse. The monitor switches to four views of den one from different angles.

Kala appears in the bottom left screen as she clambers down the giant staircase. She crosses the floor and exits the screen again, quickly reappearing on the upper level of den one, near Jock, and sits down.

'That's it,' Bob murmurs. 'Slow it down.'

McCraw reaches a hand toward one of the sagging frayed ropes that cross the den floor below.

Touni is making a huge nest of wood-wool opposite, pulling great arm loads around herself. Perfect Juni is there with her somewhere, but I can't make him out.

The percolator begins to bubble and cough around the corner in the kitchen.

'Have to keep an eye on Kera,' Bob says. 'They've all got front row seats, look.'

Kera and co are peering through the slides and dividing wall of den one, which is covered in tiny grade mesh so no baby hands or arms can get yanked through, an audience to how the mix will progress.

Jock gets to his feet, walks to the top of the steps and shifts his massive frame down the stairs to the den floor. We pick him up on the lower screen as he begins to pick at the few remaining seeds down there.

'People need to check for cockroaches,' Hannah mutters from the coffee machine. 'How fucking hard is it?'

'Protein,' Bob says.

Kala struts along the upper level, with McCraw still clinging to her belly, and sits next to Touni on the glass in reaching distance of her giant nest.

'Here we go,' I say loudly. Louder than I expected, and I know I've shown them both my exasperation. I point at the monitor.

'Can they see each other?' Bob stands at the desk to squint as close to the computer screen as he can get. 'Look at Perfect Juni. They're looking at each other, aren't they?'

'You what?' Hannah's soaked feet slap as she abandons the coffee machine to come and look up at the monitor.

McCraw puts out a tentative spidery thin arm.

'He's trying to touch him, look.' Bob snatches up the clipboard and makes a note on his scribbly mixing chart.

I hold my breath.

Touni brushes off McCraw's arm and turns her own son away from him, shuffles around so they can't see each other.

'Miserable cow,' Bob mutters.

Kala grabs a massive armful of Touni's nest and flounces off with it and McCraw clings to her in sudden panic.

'Calm down,' Bob hisses, and I can't tell if he's talking to himself, us or the gorillas. 'Pair of bloody idiots.'

Hannah glances at me with her raised eyebrows.

Touni glares, as Kala proceeds to drop the stolen wood-wool bedding off the upper level. A great bundle of it drops to the floor of den one and brushes Jock's back.

I see him jump as I glance at the lower screen, and he glowers up at the females. His chest reverberates as he coughs at them, not that we hear him from out here.

'You bloody calm down an' all,' Bob says.

The smell of coffee mixes with the damp stench of Hannah's socks.

'McCraw looks knackered already,' she says. 'Why can't she just sit down, let him rest.'

'Like this all day yesterday,' Bob mutters. 'And when he does fall asleep, she just pisses off and leaves him.'

Hannah raises her eyebrows at me again. Bob is talking like he's forgotten the three of us were right here yesterday and all witnessed the same mix together.

Kala stalks back to sit near Touni.

Jock continues to glare up at them.

'Let's put the browse in now,' Bob says. 'Four sugars when I'm back.' He hands me the clipboard.

I mark the time on the mixing sheet. *16.10. Browse added.*

Bob clatters off downstairs. We hear the grumbles of approval as he opens the door to the keeper-corridor. He keeps it quick so McCraw doesn't abandon his mum, thinking we're getting him out early.

I get a close-up of Bob's masked face, his ear pulled forward by the elastic loop. He stands in front of the camera and thrusts long, leafy branches through the mesh keeper-window.

Jock takes an armload. Kala reaches down from the upper level, squashes McCraw beneath her to swing a long arm down and grabs as many branches as she can get away with.

Touni sways her way across the ropes, above Jock's head, Perfect Juni with his arms and legs clamped around her forearm. She also collects an armful of browse and wobbles her way back across the ropes and webbing with Perfect Juni still clinging on just as he should be and, with the browse clenched in her hands and mouth, she returns to the glass panels above the public area.

They all settle down to eat.

Bob's face re-crosses the screen again and he pauses to give us a wink and a thumbs up sign. We hear the door swing open and contented grumbles from the gorillas follow him out as he clangs it shut behind him.

Kala rejoins Touni again, but they ignore each other, both content stripping leaves.

'He's grabbed a leaf, look.' Hannah points.

McCraw clamps a leaf between his teeth, but Kala is too quick for him and tugs it out, to jam in her own churning mouth, leaving him empty-handed.

Perfect Juni leans into view, his small body behind Touni. He cranes his head to catch another glimpse of McCraw.

Touni jams a tear of bark in her teeth and peels off the entire length, jerks the whole stick clean, in swift, precise tugs of her jaw. She lets Perfect Juni fiddle with the stripped white branches. He presses them to his lips and looks McCraw's way again.

Bob clumps back up the stairs. 'I'll get the coffees then, shall I?'

I write down *16.23. Grabs leaves. Kala steals.*

'It's there,' Hannah growls. 'On the side.'

The browse keeps them quiet while it lasts. McCraw totters away from Kala and picks up handfuls of wood-wool to hold above his head.

'Touni's watching him.' Hannah blows on her coffee. The mug has old coffee dribbles all down the peeling red panda logo.

'Perfect Juni's trying to get to him,' Bob whispers. 'Mark it down.'

I jot away on the sheet.

Touni shifts her son away again.

'Bloody fun-police now,' Bob snarls. He's watching the wall monitor with us now, hands on hips, frown on his face.

Kala rolls on her back and pulls McCraw up onto her chest. She begins to pat him repeatedly on the back, way too heavy handed as usual. His little body shudders and he tries to crawl off.

'She's laughing, look.' Bob shakes his head. 'Gently,' he snaps at the monitor. 'No wonder he don't like it, what does she think? He's choking on something?'

Still on her back, Kala lifts McCraw up with her arms and feet and jiggles him around and his little body bounces and shakes. He makes a concerted effort to wriggle free and staggers off toward Touni.

She waits for him to get in range and then, immediately and deliberately, she pushes McCraw off the edge of the glass hide.

He cartwheels. Drops the 3 metres to the den floor below.

I gape inside my mask.

The lower screen on the monitor picks him up. He wobbles to his feet, his lips puckered into hoots.

'Fuck,' Bob hisses.

'Bitch,' Hannah gasps.

Jock coughs, we see his throat and chest vibrate.

Kala dithers at the top of the stairs on the screen above.

McCraw begins an arduous climb back to his mum, a scramble, each of the giant steps a mammoth effort. My palms twitch, tingle at the memory of supporting his back as I helped him scale the climbing frame, or the sofa. To climb each step he has to make a grab with an arm, one back foot scraping for purchase, and he momentarily disappears from the monitor before he tumbles back down the stairs and into view. He must have rolled in piss on the step above, his fur is wet and spiked.

Jock ignores him, sits with his eye on Kala and Touni above.

McCraw goes off screen again, pulling his way up the first step and out of sight.

Kala reaches down her arm as soon as she can without the risk of Jock grabbing at her and scoops up McCraw. He clings on and she returns to her spot next to Touni.

We wait, coffees forgotten, and watch as McCraw forces his way free of Kala's heavy patting. Again, he takes a step toward Touni and she waits, eyes on Kala.

Kala isn't paying attention as she is too busy looking at Jock.

Touni times her moment and shoots out a hand to grab McCraw's leg, yanks him off his feet and spins him across the top of the glass panels.

He tumbles, Kala bounds after him and even through the closed door, we hear them cough and scream. A ball of rolling gorillas above the public area. Jock barges up the stairs, bursts into view on the main screen and both females retreat to separate corners. McCraw scurries after his mum and he's been dragged through one of the female's stress-shit, all up his back.

I stare in horror at the screen.

'Okay,' Bob says. 'Jock's sorted that one out.' He's chewing the inside of his cheek, blinks back tears. 'Poor little mite,' he whispers.

The females sit and scowl at each other. Jock lumbers back down to the den floor again and picks up the wood-wool Kala dropped earlier, shoves it into a pile to rest his head on and lies down.

Touni lets her son wave his arms at McCraw.

Kala waits for Jock to get comfy and stalks back to sit near Touni.

'Why does she keep taking him there?' Hannah rubs her eyes.

'He's well tired,' Bob says. 'Climbing them stairs.'

'Is Touni trying to use Juni to lure him over?' Hannah's voice is tight in her throat.

'Stop patting him so bloody hard.' Bob shakes his hands at the monitor. 'Oh shit, here we go.'

McCraw has clambered off Kala again.

'She's going to do it again,' Hannah says.

Touni waits, her eyes dart from Jock, to Kala, to McCraw and she shoves him. He tries to cling onto the edge but drops to the den floor again.

'I can't fucking watch this,' Bob says. 'Can't see him at all now, where is he?'

McCraw appears, fur clumped and damp, his lips puckered in hoots of fear and begins his clamber back up the stairs again. He's slower than last time.

'Is he favouring that arm?' Hannah says. 'His left?'

I keep silent, scrawl the mix notes. If I stop, I'm going to lose it. 'We have to get him out,' I blurt. It vomits out.

Bob nods. 'Nothing positive about this.'

Hannah jumps up and pulls her boots off the radiator.

Kala reaches down as before and pulls McCraw to her. She stands him on the edge of the upper level herself, eyes on Touni.

Touni ignores her.

Kala dangles McCraw at arm's length. He grips her wrist with his tired arms and his eyes dart around for anyone else.

Touni watches.

Perfect Juni is fast asleep.

Kala jiggles him over the edge. Holds her own squirming son in one hand and throws him off. His arms flail as he falls.

We all switch to look at the lower screen and he bounces into view. He staggers in bewildered panic and he makes for Jock. The huge mass of muscles, 20 times his size, his dad, protector of the family, silver fur gleaming on the screen.

McCraw wobbles toward him, lips hooting, and Jock bats him away. One huge swipe of his hairy hand sends him sprawling. He tries again, darting looks left and right in panic, and Jock leans over him, engulfs him beneath his massive shoulders, bulging arms, and leans right over him to pin him to the floor with his teeth.

I freeze.

Hannah and Bob are rigid with fear.

McCraw squirms.

Jock keeps him pinned for a few moments, then releases him. McCraw scurries in a circle and drops in the corner of the enclosure.

'Zoom in,' Hannah shrieks. Bob spins to the computer and clatters with the mouse.

The image enlarges on the bottom screen and I see him curled up, with his knees to his chest and eyes squeezed shut. His little body shut down at the horror of nowhere to go, with no one to turn to that will protect him.

Cold coffee slops. Hannah abandons her boots. Bob's chair goes over as we scramble downstairs to get him out.

Hannah operates the slides. They clunk and grind as she shuts down the isolation den.

I unlock the padlock and slide open the keeper-door.

Both Kala and Touni peer down but don't come for McCraw as Jock's simmering menace rolls around the enclosure.

McCraw sits bolt upright at the sound of the slides and hoots as he sees me.

Hannah buzzes the slide open to McCraw width and he's on his feet but cowers at the sight of Jock in his path.

I hear Kala running above us as I step into the den and crouch down.

Bob shakes a tub of peanuts for Jock at the keeper-window and McCraw darts across the floor, his little legs pumping, as he scurries past Jock and into the isolation den.

I scoop him up and feel his heart pound against my chest. 'I'm sorry,' I say. 'You're okay. I've got you.'

Chapter 20

Bath

I leave the traumatized McCraw upstairs with Hannah in the messroom and head for the flat, step out into the fading afternoon light, beneath heavy bulbous clouds and peel off my mask. My sweat freezes in the icy wind and I feel the longed-for sense of relief, that another day of trauma for McCraw is over. Why are we putting him through this?

The gibbons perform their end of day final whooping duet, and the repeated grunt roars of the lions echo across the empty zoo as I reach the flat and yank open the door. There's an odd number of wellies in the hall – how that's happened I'm not quite sure. I feel the heated warmth of the flat as I plod upstairs.

This last week has given me a real battering, a physical feeling of lethargy, and morally I feel broken.

I let the hot water blast across my head and back in the shower. We're almost out of shower gel, the bottle squeezed and full of large hexagonal bubbles.

'You poor little thing,' I mutter, like he's here instead of over in the messroom with Hannah. 'That was the worst day so far,' I add as I turn off the shower and avoid looking at myself in the steamed-up mirror as I pad straight to the bedroom. Back to the routine, block out the images of him cartwheeling down to the den floor.

Make the bed.

And Kala, his own mother, like they were using him to show off to Jock or something? Jock. Wanting nothing to do with him, too small for him to comprehend, what was that about, pinning him like that?

I catch sight of my reflection in the window as I step into the kitchen. I look weird, like a crazy person, bags under my eyes, worse than usual, sort of hunched and deflated.

'That was a hideous fucking day,' I tell myself.

McCraw's face pulled in fear and then just shutting down. Asleep in the corner, motionless, playing dead? The desperate scuttle across the floor of den one when we went in to get him out.

An unexpected cry rolls out, gulps up my throat. My reflection looks shocked. Pull it back. We're not doing that to him again. Mixing him with Kala is over.

I scan the kitchen one last time for grab hazards. Yesterday's knives and forks look precarious in the plastic drying rack. Bob's plastic takeaway cartons from last night. Coffee mug from this morning and chipped white plate, still brushed with Bob's toast crumbs.

I pack it all away, grab the rubbish bag as I leave and drop it in the bin on my way back to Gorillas and McCraw.

Hannah hasn't moved, still sitting with him on her lap upstairs in the messroom, her feet up on the wonky chair with no back. She's exhausted, beyond tired, but not ready to let go of him just yet.

I take up my spot in the corner and pick the clipboard off the floor, wet and smudged with coffee. The last thing I noted down on the mixing sheet was *Touni – cruel*, underlined repeatedly.

'Cruel' is of course a word that carries weight; we as humans know what it means. When a person is cruel you recognize that behaviour, but here I am, as a zookeeper, again ascribing a human term to animal behaviour. Something in me broke as I watched how they treated McCraw.

'That's the end of that,' Hannah says.

'Have the vets checked him over?'

She nods. 'He's going to be sore from being dragged about, but nothing's broken. There's a treatment sheet for Calpol.' She frowns. 'I just can't take it.'

'Nothing could make me put him back in there tomorrow.' I don't care that I've lost my professional perspective. 'I won't subject an animal to that level of despair.'

'It's like I'm this spring.' Hannah doesn't look at me, she peers off into space. 'Like coiled around in a clock and if one of those coils slips, the whole thing is just going to burst apart.'

'It's way harder than Afia,' I say.

Hannah glances at me. 'Afia was a walk in the park, wasn't she?'

'Remember how terrified we were when we left her with Romina?'

Hannah snorts. 'Little did we know.'

'It's the whole single parent thing on top I reckon.'

'How do you mean?'

'Why we've taken such a hammering emotionally. This time round it's way more intense.'

'And the fact life has gone nuts in general,' she says and plucks at her mask.

'With Afia, she was coming home with us, right, so even though we were doing all the caring, there were people around, family. Moral support, conversation.' I hear myself repeating my mantra.

Hannah nods. 'It's lonely in that flat.'

'How do single mums cope?' I say. 'And we get nights off. Can you imagine going through the whole thing on your own?'

McCraw shifts in her arms and clops his lips together.

'I know we have to get him in a group,' Hannah says. 'He's a social primate, but doing that again... tomorrow. No way.'

The messroom lights go off, as neither me, Hannah nor McCraw have triggered the motion sensor for 10 minutes. We're bathed in the blue light of the gorilla camera monitors. The group are mooching through each other's feeding dens, picking over the usual leftovers in a slow circular forage.

Kera waddles along, sideways, only using one of her arms, moving like she's leaning on a crutch, so she can carry a selection of leftover lettuces and cucumbers clamped to her belly with her other hand. She glances up at the camera as she passes and for the briefest moment her big face fills the screen.

McCraw jerks awake on Hannah's chest and lifts a bleary face.

The messroom lights come on and we see his shiny pink gums as he yawns.

'Let's go, you,' I say to him.

Hannah gets up and he reaches his arms for me.

I take him and he clings onto my string vest. He stinks of gorilla stress shit, dried and matted in the fur on his back.

'I've packed the shampoo,' Hannah says. 'It's in with the clipboard.'

We hear the clink of keys and Bob clatters up the stairs to join us. 'Well, that isn't fucking working, is it,' he says. 'Did he have any injuries?'

Hannah shakes her head. 'Not sure about his leg, where Touni dragged him. He's using the toes to grip okay.'

McCraw is fast asleep again, his skinny arms flung about me.

'Just spoke to the Higher-ups,' Bob says. 'Meeting later tonight, vets, curators, everyone.'

Bob often takes a leading role with the Higher-ups, even though technically Hannah and myself outrank him, but when the shit goes down, none of us care who is the more senior.

We both nod.

'Got time to get something down on paper, Al? Before you take him back to the flat.'

Tears prickle in my eyes. 'We can't put him through that again.'

Bob nods. 'Nothing to be gained, he ain't safe.'

'Think about the impact it's having on him long term too,' Hannah says. 'We don't want to create a massive silverback who is petrified of females. He'll end up killing them – look at how Jock used to be with Kera.'

The chair creaks as Bob sits down. 'Right.' He picks up the clipboard and attaches a sheet of scrap paper to it, rests it on his knee. 'One: there are no positive behaviours being shown, or any positive behaviours are quickly outweighed by all the shit we've just seen.'

Hannah nods.

'We have exhausted the mixing him back with his mum option,' I say.

Bob scrawls away on the clipboard and pauses. 'You suggesting we keep hand-rearing him until he's older, though?'

'It's an option.' Hannah looks at me. 'Right?'

'We know the older he gets, the more imprinted he'll become,' I say.

'His best option is to go with a surrogate,' Hannah states. 'There was one at another zoo wasn't there? The right age.'

'That would be the best option,' I agree. 'But it would need to happen quickly.'

Hannah sighs.

'I'm going to take him back to the flat,' I say. 'Give him a bath.'

Bob nods. 'Alright, mate.' He stands up as I step past him and reaches out a big hand, his fingernails grimy. 'You okay? You been real quiet today.'

'I'm alright. Let me know what they say at the meeting later.'

He squeezes my shoulder.

Back at the flat I flick on the kitchen light and McCraw's toes grip my belt. He makes a half-hearted reach for the Paignton Zoo elephant fridge magnet.

I turn on the tap, let the water splash into the deep aluminium kitchen sink and he shuffles up my back to peer over my shoulder. There are more toast crumbs and a torn strip of slimy lettuce waggles around the plughole spokes.

'Where's the plug?'

I crouch to open the cupboard beneath the sink.

McCraw reaches over my shoulder and flexes his fingers at a bin bag, but with no intention of grabbing it. The plug is there, a long black rubber tubular thing.

He watches the water fill the sink and looks at our reflections in the kitchen window.

It's fully dark outside now. There are three bins lined up by the muck-trailer that someone's forgotten to tip. Snowflakes flurry in the wind, a shade of brown in the orange light cast from the streetlamp outside the zoo.

I pluck a blue disposable glove out of the cardboard box, shift him away from the sideboard as he sits on my hip and pull it on. He's definitely not himself. He'd try and stuff the glove in his gob, at least try to pat at the box. I tug on a second glove.

'I'm so sorry,' I say to him. 'That's not happening again.'

I see my reflection talking.

He scrambles around me as I lower an elbow into the water to check the temperature and pour in some shampoo. Add some more warm water and wait before pulling off his nappy. He hasn't shat it in, happily. Wonder if he'll shit in the sink?

He freezes when I lower him in. The water slurps up across his belly and he sits perfectly still. Doesn't panic, but his little hands grip my vest.

'Shampoo time.'

His nostrils flare as I rub shampoo into his fur and lather him up. I squint at the shampoo bottle, something in the directions about not rinsing it off for 5 minutes.

He remains still, but his grip relaxes.

'Another minute or two,' I say.

He shifts his weight and lets one of his feet poke out above the bubbles.

I begin to rinse him off, using Bob's plastic takeaway carton as a scoop, but I can smell the shit still and have to shampoo him up for a second time.

He gurgles when I hit the tickly spot under his ear.

'Can't believe how well you're doing,' I say.

He nudges my finger with his chin and tilts his neck.

I tickle him and he laughs. Lets go of my vest with one hand to gently lower his fingers into the water.

I continue to rinse him off, tubs of water, one after the other.

He slaps the surface with both hands and slips. Grabs me with a wet fist, his arm all stringy-looking, the fur clinging to his grey skin.

I lean over him and support his back, rinse some more and he pulls the rubber plug bung thing out and bites the end.

'Bath-time over, is it?' I say. 'You don't smell anymore at least.'

He grins up at me with his play face and I pluck his drenched spidery body out of the sink and take him through into the lounge. The towel waits on the floor, in front of the warm radiator. I crouch and bundle him in it and he cackles and thrashes as I try and dry him, rolls and wrestles.

All his limbs are moving fine, he's not tender in any of them. Fuck knows how, with the way Touni dragged him about and the repeated plunges off the upper level. I know they're tough, way more robust than humans, and much of gorilla physical interaction is rough. I can't shake the image of Kala chucking him off the edge, though. Touni I can understand – I resent her for it, but I get it. Kala mistreating him was the end. Proved he wouldn't be safe with her as his mum. I reach for a logical explanation, why the two of them were being so horrible to him, even when Kala's maternal instinct remains so strong, but I can't settle on anything.

His fur stands on end, all fluffy.

'You look like a mountain gorilla,' I tell him, and he bustles off out of the lounge to go and check himself out in front of the mirror in Hannah's room, or as he's still convinced, go and look for the other gorilla that lives in this flat.

I follow him, on hands and knees. Gorilla grumble as we go. He pauses when he sees his image in the mirror. Pushes his chest forward, curves his back and marches forward, darts his head behind the mirror in the hope he'll catch the other gorilla at last. I disentangle him as he tries to climb up the mirror frame.

'Let's get your toys out.' I shift him onto my back and crawl back into the lounge. My knees are killing and happily he scrambles off me, gaze fixed on his toy box.

I turn on the TV as he upends the crate and scatters his toys across the floor. The empty chocolate tub, lid on, rolls away and circles off behind the couch.

He holds my vest and stands up on his legs to peer after it, wary of its apparent ability to move on its own.

He's completely back to normal. Having fun even, laughing and rolling about. All memory of his ordeal seemingly forgotten. The trauma of putting a human baby through all that, a toddler I suppose would be the equivalent, would surely scar them forever, but here McCraw is happy as ever. Here with us in the flat.

Eventually he runs out of steam and I give him his bottle. He drinks it slow and steady, eyes heavy-lidded toward the end, and I take him back to the couch and get comfy so we can lie back against the cushions. I treat myself to a clean face mask, smelling of laundry conditioner.

McCraw slumps across me and his grip on my arm relaxes, his fingers twitch and he sighs.

If we're going to hand-rear him until he's strong enough to look after himself, we'd be talking years. Three years, maybe four, and at some point he'd become too strong to take home. He'd live in the gorilla house, separated from everyone else overnight when we weren't around to oversee him. We work 9 hours a day and of those we'd be able to supervise about 6 or 7. That would mean for over two-thirds of his time he'd be on his own and isolated. That's where the problems would arise. Time alone.

I yawn and my phone buzzes, three messages which come in one after the other.

All three messages are from Hannah.
'no surrogates'
'the one we were thinking of has got pregnant'
'we're going to try Kera'
McCraw lifts his head.
'Kera,' I say to him. 'She's our last hope.'

Chapter 21

Kera Is Our Only Hope

I said at the beginning this was a book about Kera.

Kera is enormous in comparison to the other females. Broad shouldered and strong. She's clever and responds well to training, in part because she's had a deeper and more consistent relationship with us as keepers than she ever had within the group.

As a result, she is also less concerned about being shut away from the rest of the gorillas than anyone else. With Kala and Touni both bad news for McCraw, we begin to train Kera to be his surrogate mum.

We teach her to put a cuddly toy in a crate, like we did with Romina when we were about to introduce Afia. The idea being, as before, that if McCraw gets horribly injured and is incapable of coming to us himself, she will deposit him in a box.

We work on their relationship through the mesh and get to the stage where he is toddling in with her.

I know this is his last chance. There are no other suitable surrogates across Europe, so if we fail to get Kera to take him, he'd need to be hand-reared for a couple of years and would miss out on the foundations of gorilla social society. We'd be repeating history, creating an animal that would struggle to fit in and the heartbreaking reality would mean he would spend most of his time alone.

I force my emotion out of the mix, as I can't risk Kera or McCraw picking up on it and skewing their interactions.

Kera grunts as McCraw toddles in with her for the first time. He doesn't respond.

Bob, Hannah and I hold our breath.

McCraw potters about, sits in a pile of wood-wool and Kera eventually moves off into a different den.

We're mixing them in three interconnected feeding dens upstairs, with the rest of the group locked away in den two.

Kera's indifference was not quite the initial interaction we were hoping for but was a huge relief all the same. Kera was respectful of him.

We increase the time they spend together over the next week. Kera brings McCraw to us when we ask her to, but that's it in terms of physical contact. We need to force the maternal bond and so we leave them together overnight.

Each morning we trawl back through the grainy CCTV footage, the cameras all in monochrome night mode.

Once the lights go down, McCraw searches the dens for Kera, lost and alone in the darkness. If he bumps into her or comes into the same den she moves away and makes a nest on a platform, where she sleeps out of his reach. Eventually he lies down, a tiny patch of black fur on the screen, and sleeps alone.

On night four everything changes.

Bob and I watch back the recorded footage and McCraw begins his usual creep about in the dark. Kera had chosen to build a wood-wool nest on the floor and we watched the black and white footage of McCraw as he felt his way forward to touch her leg.

Kera jumped, but rather than move away as normal, instead we see her reach down with a long hand and scoop McCraw up.

'Fucking yes,' Bob whispers. 'Kera, you absolute legend.'

They spend the night snuggled up together.

Their bond develops, but even Kera is beginning to get stressed being separated from everyone else. She lets McCraw ride on her back, but every so often his fingers clutch too hard and she peels him off. Over the following week we let her take McCraw out into den three, rewarding her for carrying him and bringing him to us for milk feeds.

It is time to start mixing them with the rest of the gorillas.

Afia has picked up the mantle of dangerous juvenile, like Moki and Kukena before her. Seeing McCraw getting what she remembers as her bottle still drives her mad with jealousy and she has the full support of Jock in any ensuing interactions. For months we have to separate Kera and McCraw from everyone else overnight to keep him safe.

The group are all respectful of Kera, however, mainly due to her strength and size. If she forgets McCraw, we ask her to go and get him and generally she does. Interestingly, when Kera abandons him to dash outside for breakfast, Kala, his actual mother, takes a turn. McCraw is happy to hang out with his mum but always returns to Kera for comfort and safety.

Jock is getting elderly. In the wild he'd be dead by now. Having no access to veterinary dentistry, he'd have starved to death with fractured teeth, or been horrifically injured by a rival male. He has lost his canines so is no longer able to inflict the savage bites of old. Fights in the gorilla house are rarer these days, but Touni sees the opportunity to manipulate an attack on Kera by trying to get hold of McCraw. Kera goes for her. No messing about. Attacks Touni and as Jock barges in, rather than flee and abandon McCraw, Kera holds her ground and shields him with her body, lets the enraged Jock beat her back and bite at her with his gums. Kera puts her body in harm's way to protect McCraw and it is clear the instinctive maternal bond is formed.

With McCraw safely in the group I return to lead the small mammal team and try and inspire them to take up the challenge of rehoming all the species we look after.

The zoo closes its gates to the public for the final time and my own redundancy date is fast approaching. The months speed up the closer it gets. Piles of paperwork, finding suitable transport crates for meerkats and naked mole-rats, and each species we successfully rehome adds a weight of gloom across us all as we send animals we'd all worked with for years off to other zoos. Keepers begin to disperse too and Twilight Scott leaves for a permanent job with the gardens department.

I can offset my own gloom each day by feeding McCraw, or Hasani as we now collectively call him. I get in early to feed him his morning bottle, before heading off to Twilight World for the day. Hasani has reinhabited his gorilla name as he joined the group and I leave calling him Wriggles McCraw behind. He is now 20 months old and has been in with Kera and the rest of the group for just over a year. Like human babies, apes develop more slowly than other primate species. He is still small, six times smaller than Kera, but is sturdy and mobile, desperate to play with Perfect Juni when his mum Trim Touni allows it, and he remains nervous of Afia. However, he has support in Kera and his mum Kala, who acts like an older sibling, and the group all remain mixed together day and night.

I arrive at the gorilla house for the final time.

Four o'clock on my last day as a zookeeper. It has finally arrived.

The smell of coffee reaches me as I trudge upstairs to the messroom.

'One last cockroach coffee?' Hannah asks.

'Great.'

She gives me a sad look. 'I wish you hadn't gone for VR.'

Voluntary redundancy. I know part of my decision was, as Sparky had warned me, way back when during our lockdown Christmas, cutting my nose off to spite my face. The reality of no longer seeing these animals on a daily basis grips me. The gorillas or any of the other animals after 14 years. I feel a lump in my throat.

'How are you feeling?' Hannah asks, and she has tears in her eyes too.

'Not great.'

'You can come in and see them all anytime,' she says. 'This doesn't have to be it.'

She's right, for as long as I stay in Bristol that is, but I know the visits will become less frequent and at some point, end altogether. I won't be benefitting them in any way with my presence, just maintaining a bond for my own sake. Afia will be moving to a different zoo in the not-too-distant future too and will become a breeding female herself in another gorilla group somewhere.

'Fancy a pint later?' Hannah passes me the chipped red panda mug.

'Sure. We could go to the cat pub after work?' I know Hannah loves the cat pub. Not that the cats seem to like her, but she's always willing to give it a go. 'These last few months have flown by,' I say and take a last look around the messroom, somewhere so familiar that as of tomorrow I'll never come to again.

'I'll get his bottle ready for you,' Hannah says. 'I'm assuming you want to give him his last bottle?'

'I do.'

I slug back coffee as Hannah makes up the bottle and look up at the monitor.

Kera and Hasani are in the isolation den, having demolished their tea. Kera is fired up and ready to be let out. Buzzes with anxious energy and begins to roll backward and forward across the floor.

Hasani sits and watches, like he did at the flat when he was unsure.

Kera sits up and claps her hands. Shakes her mouth and when she gathers herself up for another roll, Hasani leaps in and tumbles with her. Kera ignores him, or doesn't seem to notice, and he rolls again, bats her as he goes.

She stops to stare at him.

'Look at her face.' I point at the screen and Kera is almost slap bang in front of the camera. Her face giant-sized on the monitor. Her eyes stare, eyebrow ridge raised and her lower lip dangles in what I can only describe as a smile. More than a smile, a gaping grin of delight at what she's seeing. It's a look I've never seen on her face before.

Hannah gasps at the screen, holding a hand to her mouth.

Hasani rolls again and Kera joins him, but her energy is restrained. She pokes him as he rights himself, helps him roll, and her face remains hung with delight. No way of misinterpreting the look she gives him. Joy. Sheer joy.

I don't want to interrupt them. This is the first play behaviour we've seen between the two of them and the look on Kera's face says it all. At last, after all these long years, she's getting social interaction that she enjoys. She's learned to be gentle with Hasani, but on top of that he's bonded to her and has her back. He copied her rolls, joined in with her game. He wants to spend time with her and enjoys her company.

I take a banana from the tub and go into the upstairs keeper-corridor instead, to say the first of my gorilla goodbyes.

Afia waits in the end feeding den and lets out an excited high-pitched grumble, although probably aimed more at the sight of the banana than me. We don't give them fruit as part of their diet but use it to train them or medicate them.

I crouch in front of the keeper-door and Afia slides her arm out through the gap, palm upward. Her hands are now the size of mine; she used to curl her fist around my little finger.

She takes the banana and grumbles in appreciation, slides her hand back under the keeper-door and splits the skin with her teeth. She expertly peels it, biting it into halves and gulps both down.

'I'll miss you,' I tell her. 'And I'll come and visit you wherever you eventually end up.'

Afia is as delighted with the banana skin as she was the fruit and devours the lot. She puts her hand beneath the mesh again. The truth now is that I no longer trust Afia not to grab me. I don't think she would, but I don't know, and the whole point of hand-rearing both her and Hasani in the first place is to let them be gorillas. I don't pat the top of her hand but give her a final peanut and grumble goodbyes.

Downstairs, Hasani and Kera are still playing. I creep past the isolation den to the sound of their chuckling laughter and step lightly down the keeper-corridor to the end window to say goodbye to Jock.

He waits on the training platform, rests one elbow on the windowsill and drops his huge head down to look out at me.

'Come to say goodbye, mate,' I tell him.

He shifts around to lean his shoulder against the mesh and I do the same. The mesh is between us, but we lean against each other, shoulder to shoulder. It was a behaviour he first instigated when Sal died all those years ago and he was miserable and sad. Sometimes when he sees me he comes over and does it, other times not, and I want to think he chose to today because he knows how bereft I feel. He grunts and I grumble. He doesn't register the tears rolling down my face.

'Be kind to Kera.' I sniff. 'She's looking after your son, don't forget. Doing a great job too.'

He grunts again and pokes my arm with a big fat finger.

I take a peanut out of my pocket and deposit it on his waiting lip. He splits the shell with his teeth and cups his hand on the opposite side of the mesh.

I empty my pockets of peanuts and push them through the mesh. 'Goodbye Jock,' I say.

Hannah has been keeping track of my progress on the CCTV, doesn't want to intrude, but when I come back out of the keeper-corridor to head up into the messroom for Hasani's milk, I find it waiting on the stairs with two squash bottles of fruit tea for Kera.

As far as Kera is concerned, she is getting a treat alongside Hasani having his milk. She's never tried to take the milk herself, as we've always offered her an alternative.

She rushes to the keeper-door and pushes her thick fingers through the mesh, hopeful that I'll hand her something tasty to go with the fruit tea.

'Where's your baby, Kera?' I say.

She grunts.

'Go get your baby,' I tell her.

Hasani darts about on the other side of the nesting basket, running out of Kera's range, chuckling.

Kera shuffles closer and manages to grab his wrist, tugs him over toward the keeper-door, and he does a mini grumble all of his own when he catches sight of the milk.

I offer Kera her fruit tea through the mesh with one hand and Hasani his milk with the other. He chugs it down and I time Kera's fruit tea, so Hasani's milk is finished as she gets toward the end of her second bottle.

'That's it,' I say, once they're both done. 'All finished.' I make the end-of-training-session hand signal and step away, leave Hannah to mix them all back together and bawl my eyes out as I head off to get changed.

I follow Hannah into the cat pub. 'I love this place,' she says and strips off her hat and scarf. The pub is small, narrow and full of cats.

A group of students sits in the window, a boardgame laid out on the tiny round table. They laugh and groan as two dice scatter and bounce onto the floor.

A cat launches from beneath a barstool and scatters one of the dice across the stone flagstones.

The barman is hidden from view, shielded by a row of backs and hats. The boat crowd are leant forward in a huddle, wearing thick fishermen's jumpers and rubber work boots. Trousers flecked with paint and keys that hang off their belt hoops, attached to floats. We squeeze past them to find room at the bar and a cardboard box with a tabby cat in it.

'Oh,' Hannah says in delight. 'Look at you.'

The cat sits upright in the box, tall, stretches its neck, ears flat and yawns.

We order pints and I pick a second large tabby off a bench seat to make room for Hannah.

She takes the tabby cat from my hands and sits down. Encourages it to slump on her lap. 'This will be you all the time soon,' Hannah says. 'Where are you going to get your animal fix now?'

'I'll have to get a dog.'

The cat jumps off her lap and stalks off behind the bar.

'Cheers,' Hannah says and we clink glasses.

'Here's to Kera and Hasani,' I add. 'They've figured out they can play together.'

'I'll drink to that. Kera's never looked happier.'

'Keep me updated on how they get on,' I say.

'Of course I will,' Hannah reassures me.

'And get video footage when Afia plays with Hasani.'

Hannah smiles.

'Ideally I want to see a picture of Hasani riding around on Afia's back.'

'I'll do my best,' Hannah says.

I sip my pint. 'I know we always hoped Kera's social standing would improve with the rest of the group because she had a baby.'

Hannah nods.

'But I didn't realize the joy she'd get out of their relationship, what a positive impact Hasani would have on Kera.' I slug back some more of my pint. 'That's a good memory to go out on.'

'I'd be happy with that,' Hannah says.

'I'll treasure it.'

Epilogue

The Last Supper

Bob and Twilight Scott are waiting for me at the restaurant. It's one of a whole row of eateries built in lines of shipping containers, stacked on top of each other down by the river. Always heaving, festooned with lights and high narrow tables clustered along the walkways.

Our table is upstairs, beneath a canopy that lifts and shudders in the breeze. Space heaters glow, dotted amongst the benches. I haven't seen Bob or Twilight Scott since I left the zoo six months ago.

'Here he is then,' Bob cries as I join them.

They both leap to their feet and we hug like family.

'Hannah's on a date,' Bob states. 'Probably cat bothering round the corner.' He has two empty pint glasses in front of him already.

Scott, on nights like tonight, maintains his sobriety of ten years stoically in the face of the rest of us smashing back beers and getting increasingly loud and unruly. He has a pint of Diet Coke.

'What do you mean cat bothering?' Scott asks and sits back down.

'She's obsessed with the bloody cat pub,' Bob says and belches. 'Got to be a red flag, hasn't it? On a first date?' The bench creaks as he sits next to Scott. 'She says she'll be done by nine so should be with us any minute now.'

'How's it going on gardens, Scott?' I ask and take a seat opposite them both. Scott has worked with animals his whole life and is an experienced lion keeper.

He nods. 'I like it,' he says. 'Took a while to build up my… let's say skill levels. Some of these plants are hard to recognize.'

Bob frowns. 'Still don't know why you ever went for that job.'

'I couldn't risk being made redundant,' Scott replies. 'Only ever one pay cheque away from homelessness.' Scott would often point this out and it was made all the more poignant because at one point in his life, he was indeed homeless.

'Is it satisfying, though?' Bob asks.

'You'd be surprised,' Scott says. 'And I'm still on site, but I don't work with any lions.'

Scott had trained both the Asiatic lions to receive their Covid vaccinations and modified their enclosure to create a training station. It's a memory I know he holds dear. He'd forced himself to learn the techniques of animal training, to keep pace with newer zookeepers, and was rightly proud of himself.

Unlike Bob and Hannah, Scott and I now share a reality, where neither of us will ever again experience the euphoric satisfaction of successful animal husbandry. There will be memories and stories I'll retell. Afia climbing out of her car seat, Hasani's gleeful dream, but it's going to all be memories now, for the foreseeable future at least. The gorillas' lives will go on, but I will no longer take part.

Bob salutes me with his mostly empty glass of beer. 'Can't believe you left, Al. How's your summer been?'

I'm not ready to tell Scott or Bob about my summer yet. I'll wait until Hannah arrives, so I only have to say it all once.

I shrug. 'How are the gorillas doing?' I ask him.

'Bloody great.' Bob's face lights up. 'You should see Kera, she's living the dream.'

'Drinks?'

We turn as one to a waiter with an iPad.

'What you having, Al? Pint?' Bob says.

'Great.'

'Two more of these please, mate,' Bob says and waves his glass. 'And another Diet Coke for Monty Don here.'

Scott laughs. 'Monty Don now?'

'Gardener on the telly,' Bob says. 'You can't be Twilight Scott no more.'

'What about Afia?' I ask. 'Has she got an export date yet?'

'I'll let Hannah fill you in on all that. Speak of the devil.'

Hannah winds her way between the tables.

'Hang on, mate,' Bob says to the waiter. 'Add three tequilas to that drinks order and another pint.'

'Hi.' Hannah beams.

The waiter disappears and I get up to hug her.

'It's been too long,' she says and squeezes my shoulders.

We sit back down and Hannah grins at Scott. 'Nice load of browse you delivered today,' she says. 'The gorillas absolutely loved it.'

'Always a pleasure,' Scott says.

Hannah glances up at the space heaters. 'Not very environmentally friendly,' she mutters.

'See?' Bob turns to Scott. 'Told you that'd be the first thing she said.'

Scott nods. 'You did say that.'

Hannah plucks a menu out of the holder, next to a tin can full of chopsticks and serviettes.

'Bet you been cat bothering on the way down here.' Bob winks at me.

Hannah smiles. 'We went in for one, yes.'

Bob roars with laughter and nudges Scott's elbow. 'See,' he repeats. 'Bet they all ran away.'

'Loads of this menu is gluten free,' she says.

Bob rolls his eyes. 'You're not bloody gluten free as well now are you?'

'No,' Hannah replies. 'Still just vegetarian, I was just pointing it out, that's all.'

'Fascinating,' Bob says. 'Wait… it's your new bloke, isn't it? He's a vegan?'

'How are you, Al?' Hannah asks.

'I'm good. Loads going on.'

'Never mind that,' Bob butts in. 'Let's order. He's coming back, look.'

The waiter approaches with our drinks.

'Get four portions of chicken, I reckon,' Bob says.

Hannah looks at me and Scott. 'Vegetarian? You heard me say that, right?'

'They'll be extras,' Bob says. 'One portion each isn't enough.'

'What does small plates mean?' Scott asks.

'They're like to share,' I explain.

'That's why we got to order loads of chicken,' Bob urges us. 'Have you ever had it, Scott? Korean fried chicken, like the best KFC you've ever had. I could eat five portions on me own.'

We order.

The waiter heads off to the kitchen and Bob lifts one of the shot glasses.

'Here's to you, mate,' he says to me. 'Good to see you again.'

'I'm not drinking one of those.' Hannah shakes her head.

'Course you are,' Bob argues. 'It's not for Scott, is it?'

Scott is ready to cheers with his pint of Diet Coke.

'I can't get pissed tonight,' Hannah says. 'Working. Remember?'

'We ain't seen Al in months,' Bob states. 'You got to have one at least.'

We clink glasses and throw back the hot tequila.

For years the zoo had dictated my social life, not that I minded or cared. There's no way in hell you can operate with a hangover. You only do it once. The horror of dealing with humidity, heat and gorilla shit with a booming headache is enough to ensure you never do it again.

'Why did you take VR in the end, Al?' Scott asks.

Bob and Hannah look at the table. We've never talked about my decision to leave.

'You know what it was like at the end,' I say to him. 'Up at Twilight World, all our animals leaving.'

Scott nods.

'You two could have worked carnivores out at Wildplace,' Bob says. 'Or hoofstock. Working giraffe would be alright. Head and shoulder above the rest,' he can't help but add.

'It felt like the end of an era and me and Sparky are leaving Bristol.'

'Leaving?' Bob says.

'The kids have left home,' I explain. 'We've been here over twenty years.'

The three of them look about, eyes glazed as they remember themselves 20 years ago. All still at school.

'What are you going to do?' Hannah asks.

'Something that lets me spend time closer to home,' I reply. 'The kids have gone, but my parents are getting elderly.'

'Back to the family troop,' Hannah says.

'Lucky it all kicked off after I left the zoo,' I tell them.

'You going to be a zookeeper for your parents?' Bob says. 'Is that what you mean?'

'Not exactly. But they both need a bit more input.'

I don't want to tell them any more than that.

Bob laughs. 'Let me know if you need a hand with their spot-on treatments?' He takes a huge slug from his glass.

'Wrap them in a towel,' Scott suggests.

Bob sprays beer across the table and cackles. 'Have you tried peanut butter balls? Honey water?'

'Really?' Hannah uses her exasperated tone.

'Offer them meds in sweetcorn,' Scott adds. 'The cooked kind, strip a few kernels and present it on the Bat-master 2000.'

'What's the Bat-master 2000?' Hannah asks.

'Ah,' Scott says. 'Well, if the mouse deer won't come for its meds, the meds must go to the mouse deer, on the Bat-master 2000.'

'Fragile little bastards.' Bob takes another huge swallow of beer. 'Mouse deer.'

'It was a plastic lid taped to a piece of bamboo,' I tell her. 'Scott made it.'

'Like a two-meter-long teaspoon,' he explains.

'What's that got to do with bats?' Hannah asks.

Scott concedes Hannah's query with a nod of his head. 'It's similar in design to the Bat-master 1000,' he says.

'There you go, Al,' Bob says. 'The parent 5000, that's what you need.'

Scott laughs.

'Don't encourage him, Scott,' Hannah chides.

I realize nights like this are going to become few and far between. Zookeeping conversations are going to become a thing of the past. How many more times will I see these people? Particularly when we move, and how often will I really see the gorillas?

'Bob says Kera's living the dream,' I say.

Hannah comes alight just like Bob had. 'It's been so interesting,' she replies. 'Hasani has brought Kera and Kala together, they've formed a little trio in the group.'

'You need to come in and see them,' Bob says. 'Kera's coat looks great, she ain't been plucking.'

'Which reminds me,' Hannah says and pulls her phone out of her jacket pocket. 'Wait until you see this.'

The video is taken through the keeper-door of the isolation den. Jock lies in the back half of the den, picking through the remains of an iceberg lettuce. Hasani and Afia hurtle around the front half, ducking around the nesting basket and wrestling together. Hasani darts out of view and reappears behind Jock. Afia waits for him in the nesting basket and he launches at her, his face a big grin.

'Amazing,' I say.

'And look at this one,' Hannah says and scrolls to a second video.

It's taken from the public area and shows Kera lying on her side on the upper level of den two. Hasani squats next to her and intently grooms the fur on her head.

'Nobody has ever groomed her before,' Hannah says.

Grooming for gorillas is a time of soporific relaxation and a way of reaffirming social bonds.

I get a taste of loss, at missing these meaningful moments. My life has fewer early starts now and I won't be working Christmas Day. No more taking animals up to clinic for their health checks and unloading trailer-loads of bark chip into enclosures, standing in a sweating line of keepers with shovels and wheelbarrows. No need to ever look at the staffing rota again, or holiday entitlement, an end to one-to-one meetings with my team and animal department meetings. I'll never visit the police firing range again for gun-crew training or help unload frozen horse meat for the lions and I won't be around when the gorillas eventually move to their new home a few miles up the road.

Life as I knew it has ended and there's a deep well of emptiness, but seeing Hasani grooming Kera is the boost I need. I've started a new chapter and have to leave this particular troop behind, my zookeeping gorilla family.

'When are you leaving Bristol?' Hannah asks.

'Soon,' I say.

'You'll come in and say goodbye to them all, though?'

'I will.'

Acknowledgements

This book would not have taken shape without the help of Simon Tonkin, a true friend who, over many pints, has pored over every word and discussed each chapter at length. My deepest thanks to you, for your honesty, enthusiasm and laughter.

The real hand-rearing team at Bristol Zoo – the real-life Hannahs and Bobs: Sarah Gedman, the Mighty Sam Matthews, Ryan Walker, Lynsey Bugg, Jo Rudd, Zoe Grose and Shanika Ratnayake.

The entire vet department at Bristol Zoo, all of whom had a hand in the care of both baby gorillas: Teresa Horspool, Rowena Killick, Kelly Wyatt, Richard Saunders, Nic Hayward, Sara Shopland and of course Charlotte Day, the vet that ventilated, or breathed for, Afia once she was delivered, and kept her alive until she could breathe for herself. A special thank you to Michelle Barrows, for proofreading and making sure I'd got all the veterinary procedural scenes right.

The wider Bristol Zoo family, in particular Joe Allotey who, as Kera lay on death's door, instantly procured a foam mattress for her to recover on.

Miriam Haas for the generous use of all your amazing gorillas photos. Imogen Calendar for the use of your behind-the-scenes photos.

My agent, James Spackman, for championing the book, and the editorial team at Summersdale Publishers: Debbie

Chapman, Imogen Palmer, Jasmin Burkitt and a special thanks to Ross Dickinson – the exchange of drafts and clarity through the edit was fantastic.

My own family troop of course: Sharon Clark, Seth Tims and Kiz Tims, who offered their unwavering support, not just with the book, but also with the hand-rearing of Afia. A special thanks to Sharon, for her baby knowledge and acceptance that we were suddenly sharing our home with a gorilla. I will never forget the calm, resigned expression on your face as you held a wriggling Afia over the bath as she splattered us with gorilla diarrhoea. My brother Robin Toyne and dad Derek Toyne, for their precise insights and ideas for the book.

And lastly, Kera.

The last time I saw Kera was months after I'd left the zoo. I went in to visit, and both Kera and Hasani were sitting at the mesh in the upstairs keeper-corridor. Hasani still recognized me and shoved his arms under the door to pat my hands and Kera coughed. Whether she recognized me or not, she still gave the noise a gorilla makes as a warning to back off. Kera was protective of Hasani and she has my endless gratitude for that.

Have you enjoyed this book?
If so, why not write a review on your favourite website?

If you're interested in finding out more about our
books, find us on Facebook at **Summersdale Publishers,**
on Twitter/X at **@Summersdale** and on Instagram and
TikTok at **@summersdalebooks** and get in touch.
We'd love to hear from you!

Thanks very much for buying this Summersdale book.

www.summersdale.com

Margaret Oliphant

AGNES

Volume 2

Elibron Classics
www.elibron.com

Elibron Classics series.

© 2005 Adamant Media Corporation.

ISBN 1-4021-7382-2 (paperback)
ISBN 1-4021-2485-6 (hardcover)

This Elibron Classics Replica Edition is an unabridged facsimile
of the edition published in 1865 by Bernhard Tauchnitz,
Leipzig.

Elibron and Elibron Classics are trademarks of
Adamant Media Corporation. All rights reserved.

Margaret Oliphant

AGNES

Volume 2

Elibron Classics
www.elibron.com

Elibron Classics series.

© 2005 Adamant Media Corporation.

ISBN 1-4021-7382-2 (paperback)
ISBN 1-4021-2485-6 (hardcover)

This Elibron Classics Replica Edition is an unabridged facsimile
of the edition published in 1865 by Bernhard Tauchnitz,
Leipzig.

Elibron and Elibron Classics are trademarks of
Adamant Media Corporation. All rights reserved.